STUDENT

SELF-STUDY

GUIDE

**Mortimer**

# CHEMISTRY
## A Conceptual Approach

**FOURTH EDITION**

# STUDENT
# SELF-STUDY
# GUIDE

## Donald W. Shive
## Louise E. Shive

*MUHLENBERG COLLEGE*

**D. VAN NOSTRAND COMPANY**
New York   Cincinnati   Toronto
London   Melbourne

D. Van Nostrand Regional Offices:
New York                 Cincinnati

D. Van Nostrand Company International Offices:
London         Toronto         Melbourne

Copyright © 1979 by Litton Educational Publishing, Inc.

ISBN: 0-442-25548-9

Published by D. Van Nostrand Company
135 West 50th Street, New York, N.Y. 10020

10 9 8 7 6 5 4 3 2 1

# Preface

The *Self-Study Guide* is designed to accompany the Fourth
Edition of *Chemistry: A Conceptual Approach,* by Charles E.
Mortimer. The organization of the guide follows that of the
text, each chapter corresponding to one in the text. The
guide may be used with little or no assistance from the instructor. It should be used, however, after the student has
attended lectures and read the text.

Each chapter is divided into four parts: Objectives, Exercises, Answers to Exercises, and Self-Test. The first part
provides a list of explicit objectives; the second contains
verbal and mathematic exercises designed to help the student
meet these objectives.

Exercises should be done sequentially and answers checked
immediately upon completion. The answers to all exercises
and detailed solutions to most problems are given in the
third part. The answer to an exercise is generally given

in the left column and the detailed solution to the right. The solutions include many hints, tables, and figures. Text references [in brackets] are given with many answers. The last part is a self-test and should be completed after studying lecture, text, and study guide material. The test is to be completed within a prescribed time, and answers are given without comment in the back of the guide. The self-test is representative of test questions given at Muhlenburg College and is not intended to be comprehensive or indicative of the type of questions that should be on an exam.

We hope that this guide is helpful and makes the study of chemistry more enjoyable.

—Donald and Louise Shive

# Contents

# Introduction

OBJECTIVES

I. You should be able to demonstrate your knowledge of the
following terms by defining them, describing them, or
giving specific examples of them:

Celsius temperature scale [1.2]
centi- [1.2]
chemical change [1.1]
compound [1.1]
element [1.1]
energy [1.1]
Fahrenheit temperature scale [1.3]
International System of Units and SI units [1.2]
kilo- [1.2]
matter [1.1]
micro- [1.2]
milli- [1.2]
phase [1.1]
physical change [1.1]
pure  substance [1.1]

1

II. You should be able to write numbers in scientific notation.

III. You should be able to determine and work with the proper
number of significant figures.

IV. You should study the instruction manual that came with
your calculator.

**EXERCISES**    I. Write the following numbers in scientific notation:

_____ 1. 751
_____ 2. 781,000
_____ 3. 781,000.0
_____ 4. 0.050

_____ 5. 0.000745
_____ 6. one million
_____ 7. one-millionth
_____ 8. three-tenths

II. How many significant figures are contained in each of
the following numbers?

_____ 1. 2.57

_____ 2. 0.0057

_____ 3. 0.570

_____ 4. $5.7 \times 10^{-3}$

_____ 5. $2.9979 \times 10^{10}$

_____ 6. 0.0821

_____ 7. $6.022 \times 10^{23}$

_____ 8. 900,000

_____ 9. 1.00

_____ 10. 0.75

_____ 11. 1.75

_____ 12. 1.750

III. Perform the following calculations and report the answer
and the appropriate number of significant figures:

1. $6.0 + 297 + 8.75 =$

2. $7.41 + 0.02 =$

3. $6.9 + 0.001 =$

4. $182 - 99.2 =$

5. $(7.10 \times 10^{21}) + (7.10 \times 10^{20}) =$

6. $(6.4 \times 10^{-1}) - (4.21 \times 10^{-2}) =$

7. $7.1/9.64 =$

8. $(6.022 \times 10^{23})(1.70) =$

9. $(1.074 \times 10^{-4})(9.9) =$

10. $(7.45 \times 6.1)/2.45 =$

**ANSWERS TO EXERCISES**

I. Scientific notation

It is often inconvenient to work with the standard form of very large or very small numbers. For example, the multiplication involving the numbers 701,000 and 0.00000077 can easily be done incorrectly if we lose track of the zeros. An exponential notation, commonly referred to as scientific notation, is used to simplify calculations involving such numbers. The conversion from standard notation to scientific notation is quite simple, and a method for performing this conversion is summarized as follows:

(a) Move the decimal point so that there is a single digit (not zero) to its left, for example,

0.0000007⊙7

(b) Count the number of digits between the original and new decimal point positions to determine the magnitude of the exponent of 10, for example,

0.0000007⊙7
 1234567

$10^7$

(c) Determine the sign of the exponent of 10 by the direction from the original to the new decimal point position (right is negative, left is positive), for example,

0.0000007⊙7

The direction indicates a negative sign. Thus,

0.00000077 equals $7.7 \times 10^{-7}$.

Properly used, scientific notation clearly indicates the number of significant figures. When we expressed 0.00000077 in scientific notation, we retained only the two sevens. The zeros preceding the sevens were expressed by the magnitude of the exponent. Thus, the zeros are not significant. Only the two sevens are significant. We can summarize the general rules concerning zeros in a number as follows:

(a) Zeros to the left of any digit other than zero are not significant.

(b) Zeros to the right of any digit other than zero are significant when a decimal point is included in the number. If a decimal point is not included in the number, the zeros to the right of any digit other than zero may or may not be significant.

      (c) Zeros between digits other than zero are signifi-
cant.

In the examples of this and the following section, we will
see specific cases in which these general rules apply.

1. $7.51 \times 10^2$   3 significant figures

2.         The number of significant figures to which the measurement
was made is not clear in this example. The number can be
written therefore in several forms, depending upon the precision of the measurement.

  $7.81 \times 10^5$   3 significant figures

  $7.810 \times 10^5$   4 significant figures

  $7.8100 \times 10^5$  5 significant figures

  $7.81000 \times 10^5$     6 significant figures

3. $7.810000 \times 10^5$     7 significant figures

4. $5.0 \times 10^{-2}$   2 significant figures

5. $7.45 \times 10^{-4}$   3 significant figures

6. $1 \times 10^6$   When a number if written in word form, we determine the number of significant figures by writing the number in numerical
form and applying the general rules. If the number of significant figures is still unclear, we choose the lowest number. Thus, one million is 1,000,000 and has one significant
figure. Occasionally, integers are given in word form and are
meant to have an infinite number of significant figures. (See
Table 1.1.)

7. $1 \times 10^{-6}$   1 significant figure

8. $3 \times 10^{-1}$   1 significant figure

TABLE 1.1.  Significant Figures of Numbers Written in Word Form

| Number in Word Form | Numerical Form | Number of Significant Figures |
|---|---|---|
| one thousand | 1,000 | 1 |
| one and five one-hundredth | 1.05 | 3 |
| fifty-five | 55 | 2 |
| one hundred and ten | 110 | 2 |
| two thousand and twenty | 2,020 | 3 or 4 |
| five million and twenty-four | 5,000,024 | 7 |
| six million and two thousand | 6,002,000 | 4 to 7 |

1. 3

2. 2    The zeros before and after the decimal point only show the position of the decimal point. It is better to write this number in scientific notation as $5.7 \times 10^{-3}$. *Zeros to the left of any digit other than zero are not significant.*

3. 3    *Zeros to the right of any digit other than zero are significant if a decimal point is included in the number.*

4. 2    When a number if written in scientific notation, the number of significant figures is determined by the whole number. The exponent shows the position of the decimal.

5. 5

6. 3

7. 4    *Zeros between digits other than zero are significant.*

8. 1, 2, 3,    The number of significant figures depends upon the precision
   4, 5, 6    of the measurement.

9. 3    Rule (b)

10. 2    Rule (a)

11. 3

12. 4    Rule (b)

III. Calculations

1. $3.12 \times 10^2$  After the addition is performed, the answer, 311.75, is rounded off to the correct number of significant digits as determined by the smallest number of digits to the right of the decimal point in any of the numbers that are being added. The number 279 has no digits to the right of the decimal. The sum, therefore, is reported as 312, or preferably as $3.12 \times 10^2$. This procedure should be followed whenever addition or subtraction is performed.

2. 7.43

3. 6.9

4. 83    Notice that the number 82.8 is rounded off to 83. The answer has fewer significant figures than either number used in the calculation.

5. $7.81 \times 10^{21}$

6. $6.0 \times 10^{-1}$

7. 0.74        In both multiplication and division the answer is usually
               rounded off to the smallest number of significant figures
               contained in any of the numbers involved in the calcula-
               tion. This rule normally works well, and we will use it ex-
               clusively.

8. $1.02 \times 10^{24}$

9. $1.1 \times 10^{-3}$

10. $1.9 \times 10^{1}$


## SELF-TEST        Complete the test in 15 minutes.

I. Answer each of the following:

_____      1. The process of water changing into steam is called
                  (a) freezing                 (c) a physical change
                  (b) a chemical change        (d) fusion

_____      2. The number 0.070020 has how many significant figures?
                  (a) 1      (b) 2     (c) 4      (d) 5

_____      3. The number 0.070020 should be written in scientific no-
                  tation as
                  (a) $7.002 \times 10^{2}$        (c) $7.0020 \times 10^{-2}$
                  (b) $7.002 \times 10^{-2}$       (d) $7.0020 \times 10^{2}$

_____      4. The number 700 should be written in scientific notation as
                  (a) $7 \times 10^{2}$            (c) $7.00 \times 10^{2}$
                  (b) $7.0 \times 10^{2}$          (d) cannot be determined

_____      5. One microgram is equal to
                  (a) $10^{3}$ g                  (c) $10^{6}$ g
                  (b) $10^{-3}$ g                 (d) $10^{-6}$ g

_____      6. One kilometer equals
                  (a) $10^{3}$ m                  (c) $10^{6}$ m
                  (b) $10^{-3}$ m                 (d) $10^{-6}$ m

_____      7. The number 700.0 should be written in scientific notation
                  as
                  (a) $7 \times 10^{2}$            (c) $7.000 \times 10^{2}$
                  (b) $7.00 \times 10^{2}$         (d) $7.000 \times 10^{3}$

_____      8. The process of gasoline burning in an automobile cylinder
                  is called
                  (a) a physical change          (c) vaporization
                  (b) a chemical change          (d) boiling

_____      9. The sum of the number of 71.742, 6.0, and 21.3413 is
                  (a) 99.0833                     (c) 99.1
                  (b) 99.0                        (d) $1.0 \times 10^{2}$

10. The number 199.969, rounded off to three significant figures, should be written as
(a) 200                          (c) $2 \times 10^2$
(b) 199                          (d) $2.00 \times 10^2$

11. The number 0.74 has how many significant figures?
(a) 1                            (c) 3
(b) 2                            (d) 4

12. The physical state of matter that assumes the shape of its
container only within the limit of the volume that the
sample occupies is
(a) vapor                        (c) liquid
(b) gas                          (d) solid

13. SI units are
(a) units of the International System
(b) Standard International units
(c) Substituted International units
(d) Scientific International units

14. Multiplication of the number 23.6 by the number $7.50 \times 10^3$
yields
(a) $1.77 \times 10^3$           (c) $1.7 \times 10^5$
(b) $1.77 \times 10^5$           (d) $177 \times 10^3$

15. A cow produces milk from ingested foodstuffs. This process
can be called
(a) a liquefaction               (c) a physical change
(b) a chemical change            (d) a biological marvel

# Atomic Structure

OBJECTIVES

I. You should be able to demonstrate your knowledge of the following terms by defining them, describing them, or giving specific examples of them:

alpha ray [2.5]
atom [2.1, 2.5]
atomic mass unit, μ [2.8]
atomic number, A [2.8]
atomic weight [2.8]
Balmer lines [2.10]
Bohr theory [2.10]
cathode rays [2.2, 2.3]
coulomb [2.2]
diamagnetism [2.14]
electromagnetic radiation [2.9]
electron [2.2]
electronic configuration [2.14]
frequency [2.9]

gamma ray [2.5]
Hund's rule [2.14]
isotope [2.7]
mass number, Z [2.6]
neutron [2.4]
nucleus [2.5]
orbital [2.12]
paramagnetism [2.14]
Pauli exclusion principle [2.13]
periodic table [2.11]
photon [2.9]
proton [2.3]
quantum numbers [2.13]
quantum theory [2.9]
radioactivity [2.5]
spectrum [2.10]
transition element [2.15]
valence shell [2.14]
wavelength [2.9]
X ray [2.9, 2.11]

II. You should be able to determine the number of protons, neutrons, and electrons in any isotope of any element.

III. You should be able to calculate atomic weights from masses and relative abundances of isotopes.

IV. You should understand the relationship between energy and frequency, $E = h\nu$ or $E = hc/\lambda$, and be able to work problems relating to these equations.

V. You should understand the Bohr theory.

VI. You should be able to write a complete set of quantum numbers for each electron of any element.

VII. You should be able to write the electronic configuration of each element in the periodic table and of any monatomic ion. You should also be able to predict the number of unpaired electrons and the magnetic properties of each of these species.

UNITS,
SYMBOLS,
MATHEMATICS

I. You should start building your understanding of SI notation and learn some of the prefixes that are used in conjunction with these standard units. For example, the measurement of length is the meter, m. Some of the standard prefixes used with the meter are listed in Table 2.1, along with the meaning of each symbol.

TABLE 2.1  Prefixes Used in SI Notation

| Prefix | Multiplier | Length Measurements[a] | |
|--------|------------|-------------|----------------------|
| | | Name | Symbol and Equivalent |
| pico-, p | $10^-$ | picometer | 1 pm = 1 × $10^{-12}$ m |
| nano-, n | $10^-$ | nanometer | 1 nm = 1 × $10^{-9}$ m |
| micro-, μ | $10^-$ | micrometer | 1 μm = 1 × $10^{-6}$ m |
| milli-, m | $10^-$ | millimeter | 1 mm = 1 × $10^{-3}$ m |
| | 1 | meter | 1 m = 1 m |
| kilo-, k | 10 | kilometer | 1 km = 1 × $10^3$ m |
| mega-, M | 10 | megameter | 1 Mm = 1 × $10^6$ m |

[a]The centimeter (1 cm = 1 × $10^{-2}$ m) is acceptable but not preferred usage. The Angstrom (1 Å = 1 × $10^{-10}$ m) should not be used.

The same prefixes shown in Table 2.1 are used in conjunction with all other SI units. Table 2.2 lists the units for frequency (in hertz) and time (in seconds). The underscored terms are those commonly encountered in chemistry. Many of them are used in this chapter.

TABLE 2.2  Frequency and Time Units

| Prefix | Frequency | Time |
|--------|-----------|------|
| pico- | 1 phz = 1 × $10^{-12}$ hz | 1 ps = 1 × $10^{-12}$ s |
| nano- | 1 nhz = 1 × $10^{-9}$ hz | 1 ns = 1 × $10^{-9}$ s |
| micro- | 1 μhz = 1 × $10^{-6}$ hz | 1 μs = 1 × $10^{-6}$ s |
| milli- | 1 mhz = 1 × $10^{-3}$ hz | 1 ms = 1 × $10^{-3}$ s |
| | 1 hz = 1 hz | 1 s = 1 s |
| kilo- | 1 khz = 1 × $10^3$ hz | 1 ks = 1 × $10^3$ s |
| mega- | 1 Mhz = 1 × $10^6$ hz | 1 Ms = 1 × $10^6$ s |

II. Be sure you can manipulate numbers in scientific notation with your calculator so that you can keep track of the proper power of 10 as you perform calculations. If you cannot, now is the best time to review Chapter 1 in the study guide.

EXERCISES

I. Major contributions in the development of atomic theory were made by the following scientists. If your instructor emphasizes names in lecture, you should take the time to match the name of each scientist with his contribution.

_____  1. Leucippus and Democritus, 400-500 B.C.

_____  2. Dalton, 1803

_____  3. Einstein, 1920

a. determined the charge of the electron, e, with his famous oil drop experiment

b. discovered radioactivity

c. first observed positive rays, or canal rays

_____    4. Goldstein, 1886

_____    5. Rutherford, 1911

_____    6. Planck, 1900

_____    7. Chadwick, 1932

_____    8. Millikan, 1909

_____    9. Moseley, 1913

_____   10. Becquerel, 1896

_____   11. Bohr, 1913

_____   12. Mendeleev and
                   Meyer, 1869

_____   13. de Broglie, 1925

_____   14. Schrödinger, 1926

_____   15. Heisenberg, 1927

d. proposed a theory of the electronic structure of atoms based on the emission spectrum of hydrogen

e. proposed the existence of the neutron

f. presented evidence for the existence of a nucleus in an atom from alpha particle scattering

g. proposed the quantum theory of radiant energy

h. proposed that the atomic number of an element is equal to the number of units of positive charge in the nucleus

i. early Greeks who proposed the existence of the atom

j. proposed the periodic classification of the elements

k. proposed that mass and energy are related by $E = mc^2$

l. proposed that position and velocity simultaneously cannot be measured exactly

m. postulated the wave nature of the electron

n. presented the keystone to wave mechanics

o. first proposed an atomic theory

II. Match the name of the element with the atomic symbol. These symbols are part of the language of chemistry and should be mastered early in your course.

_____    1. silver          a. Ag
_____    2. hydrogen        b. Al
_____    3. sodium          c. As
_____    4. calcium         d. Au
_____    5. potassium       e. B
_____    6. boron           f. Ba
_____    7. fluorine        g. Br
_____    8. copper          h. Ca
_____    9. gold            i. Cl
_____   10. nitrogen        j. Cr
_____   11. bromine         k. Cu
_____   12. antimony        l. F
_____   13. chromium        m. H
_____   14. zinc            n. He
_____   15. arsenic         o. Hg
_____   16. tin             p. K
_____   17. chlorine        q. Mg
_____   18. helium          r. N
_____   19. tungsten        s. Na
_____   20. lead            t. Pb
_____   21. barium          u. S
_____   22. silicon         v. Sb
_____   23. magnesium       w. Si
_____   24. sulfur          x. Sn
_____   25. mercury         y. W
_____   26. aluminum        z. Zn

III. Answer each of the following with *true* or *false*. If a statement is false, correct it. Use this section as a guide for further study.

_____   1. The energy of a quantum of radiation is directly proportional to its wavelength. The propotionality constant, $h$, is Planck's constant.

_____   2. The atomic mass, $A$, is the sum of the number of neutrons and protons in the nucleus.

_____   3. The atomic number is equal to the total number of neutrons in the nucleus.

_____   4. Isotopes of an element have the same atomic weight but differ in atomic number.

_____   5. Subatomic particles include the proton, the neutron, and the electron.

6. The charge of a proton is equal to but opposite in sign from that of the electron.

7. A neutron has a positive charge.

8. The nucleus is always a combination of protons and neutrons.

9. Isotopes are atoms of different chemical reactivity but equal masses.

10. The atomic weight of an element is a weighted average of the naturally occurring isotopes of that element.

11. A diamagnetic material has unpaired electrons.

12. Electrons are distributed in atoms so that a maximum number are paired.

13. Paramagnetic materials interact with magnetic fields.

14. The frequency of a quantum of radiation in inversely proportional to its wavelength.

15. Electrons associated with an atom can exist only in discrete energy levels.

16. It is possible for an electron to have the set of quantum numbers $n = 1$, $l = 1$, $m_l = +1$, $m_s = +\frac{1}{2}$.

17. It is possible for an electron to have the set of quantum numbers $n = 2$, $l = 0$, $m_l = 0$, $m_s = +\frac{1}{2}$.

18. It is possible for an electron to have the set of quantum numbers $n = 2$, $l = 1$, $m_l = +1$, $m_s = +\frac{1}{2}$.

19. It is possible for an electron to have the set of quantum numbers $n = 4$, $l = 3$, $m_l = -2$, $m_s = +\frac{1}{2}$.

20. A $d$ subshell can contain a maximum of six electrons.

21. A noble gas is unreactive chemically because the electronic configuration of its outer shell consists of completely filled $s$ and $p$ subshells.

IV. Using the periodic table in the back of the study guide, identify the element that is described by the statement:

1. $Z = 13$

2. The atomic weight is 74.9 u.

3. The atomic number is 50.

4. The mass number is 109 and the number of neutrons is 62.

_____ 5. There are 16 electrons in the neutral atom.

_____ 6. Two isotopes exist, one with a mass of 10.013 u and one with a mass of 11.009 u.

_____ 7. There are 18 electrons in the ion that has a single negative charge, $X^-$.

_____ 8. There are 22 electrons in the ion that has two positive charges.

_____ 9. The electronic configuration is $1s^2 2s^2 2p^3$.

_____ 10. The electronic configuration is $1s^2 2s^2 2p^6 3s^2 3p^6 3d^8 4s^2$.

_____ 11. The electronic configuration is $1s^2 2s^2 2p^6 3s^2 3p^6 3d^{10} 4s^2 4p^6$.

_____ 12. The electronic configuration of the ion that has a single positive charge, $X^+$, is $1s^2 2s^2 2p^6$.

_____ 13. The electronic configuration of the ion that has a single positive charge, $X^+$, is $1s^2 2s^2 2p^6 3s^2 3p^6 3d^{10}$.

_____ 14. The quantum numbers of the last electron added according to the aufbau method are $n = 2$, $l = 1$, $m_l = 0$, $m_s = +\frac{1}{2}$.

_____ 15. The quantum numbers of the last electron added according to the aufbau method are $n = 3$, $l = 1$, $m_l = 0$, $m_s = -\frac{1}{2}$.

_____ 16. The quantum numbers of the last electron added according to the aufbau method are $n = 4$, $l = 2$, $m_l = 0$, $m_s = -\frac{1}{2}$.

V. Write the electronic configuration of the following:

_____ 1. Ca, calcium

_____ 2. $Zn^{2+}$, the zinc(II) ion

_____ 3. $Cs^+$, the cesium ion

_____ 4. Kr, krypton

_____ 5. $S^{2-}$, the sulfur atom

VI. Write the four quantum numbers of the last electron added to the element, the differentiating electron:

_____ 1. O, oxygen

_____ 2. K, potassium

_____ 3. V, vanadium

_____     4. As, arsenic

_____     5. Ni, nickel

_____     6. Au, gold

VII. In Chapter 2 of your text you have learned to calculate
atomic weights from the masses of isotopes and to determine the energy of electromagnetic radiation from either
the wavelength or frequency of that radiation. Work the
following problems to test your expertise:

1. Boron consists of two naturally occurring isotopes.
One isotope, $^{10}B$, has a mass of 10.01294 u and the
other, $^{11}B$, has a mass of 11.00931 u. If the atomic
weight of boron is 10.811, what percent of each of
the two isotopes is naturally occurring boron?

2. Chlorine consists of two naturally occurring isotopes, $^{35}Cl$ and $^{37}Cl$. The $^{35}Cl$ isotope is mor abundant than the $^{37}Cl$ isotope (75.53% vs. 24.47%). Calculate the approximate atomic weight of naturally
occurring chlorine. Assume that the mass of $^{35}Cl$ is
35.00 u and that of $^{37}Cl$ is 37.00 u.

3. Radio and TV antennae are designed so that the
length of a crossbar is approximately equal to the
wavelength of the signal received. If a crossbar is
1 mm long, what frequency in megahertz is it designed
to receive?

4. Which has the higher energy: a 1.0 nm X ray or a
$3.0 \times 10^{10}$ hz microwave?

ANSWERS TO     I. Contributions of scientists
EXERCISES
1. i     2. o     3. k     4. c     5. f     6. g     7. e
8. a     9. h     10. b     11. d     12. j     13. m
14. n     15. l

II. Names and symbols of elements

After checking your answers, write the name of the element next to the symbol:

1. Ag _____

2. H _____

3. Na _____

4. Ca _____

5. K _____

6. B _____

7. F _____

8. Cu _____

9. Au _____

10. N _____

11. Br _____

12. Sb _____

13. Cr _____

14. Zn _____

15. As _____

16. Sn _____

17. Cl _____

18. He _____

19. W _____

20. Pb _____

21. Ba _____

22. Si _____

23. Mg _____

24. S _____

25. Hg _____

26. Al _____

III. Principles of atomic structure

1. False [2.9]    Energy, $E$, is directly proportional to frequency, $\nu$.

$$E = h\nu$$

The constant of proportionality is called Planck's constant, $h$. Energy, however, is inversely proportional to wavelength, $\lambda$, since

$$E = hc/\lambda$$

in which $c$ is the speed of light.

2. True [2.6]

3. False [2.6]    The atomic number, $Z$, is the total number of protons in the nucleus of an atom.

4. False [2.7]    Isotopes have different atomic weights but the same atomic number.

5. True [2.4]

6. True [2.3]

7. False [2.4]    The word *neutron* is thought to be derived from the same word as *neutral*. The neutron has no charge.

8. False [2.5]    The hydrogen nucleus contains only a proton. All other nuclei contain both neutrons and protons.

9. False [2.7]    Although the masses of isotopes differ, the chemical reactivities are the same. Isotopes are atoms of the same element.

10. True [2.8]

11. False [2.14]    In diamagnetic materials all electrons are paired.

12. False [2.14]    The number of electrons that are unpaired is a maximum, as stated in Hund's rule of maximum multiplicity.

13. True [2.14]

14. True [2.9]    The relationship between frequency and wavelength is

$$\nu = c/\lambda$$

in which $\nu$ is the frequency in hertz or $sec^{-1}$; $\lambda$ is the wavelength of cm; and $c$ is the speed of light in a vacuum, in units of cm $sec^{-1}$.

15. True [2.10]

16. False [2.13]    If $n = 1$, $l$ can only be zero.

17. True [2.13]    For $n = 2$, the following sets of quantum numbers are possible:

$$l = 0, \; m_l = 0, \; m_s = +\tfrac{1}{2} \text{ or } -\tfrac{1}{2}$$
$$l = 1, \; m_l = +1, \; m_s = +\tfrac{1}{2} \text{ or } -\tfrac{1}{2}$$
$$l = 1, \; m_l = 0, \; m_s = +\tfrac{1}{2} \text{ or } -\tfrac{1}{2}$$
$$l = 1, \; m_l = -1, \; m_s = +\tfrac{1}{2} \text{ or } -\tfrac{1}{2}$$

In general, for any value of $n$ the possible values of $l$ and $m_l$ are

$$l = (n - 1), \; (n - 2), \; \ldots, \; 0$$
$$m_l = +1, \; +(l - 1), \; \ldots, \; 0, \; \ldots, \; (l - 1), \; -l$$

18. True [2.13]

19. True [2.13]

20. False [2.13]    A maximum of 10 electrons can be added to a $d$ subshell. The $d$ orbitals are being filled in each series of transition elements. The possible quantum numbers of electrons in $d$ orbitals are

$$l = 2; \; m_l = +2, \; +1, \; 0, \; -1, \text{ or } -2; \; m_s = +\tfrac{1}{2} \text{ or } -\tfrac{1}{2}$$

21. True [2.16]

IV. Identification of elements

1. Al, aluminum [2.11]

2. As, arsenic [2.8]

3. Sn, tin [2.11]    The symbol for atomic number is $Z$.

4. Ag, silver
   [2.6, 2.11]

The atomic number is obtained by subtracting the number of neutrons from the mass number, 109 − 62 = 47. The atomic number is 47.

5. S, sulfur
   [2.6, 2.11]

The number of electrons in a neutral atom equals the number of protons in that atom; therefore, $Z = 16$.

6. B, boron
   [2.7, 2.8]

Boron is the only element that has an atomic weight lying between 10.013 u and 11.009 u.

7. Cl, chlorine
   [2.11, 2.15]

The ion has one more electron than the neutral atom; therefore, the neutral atom has 17 electrons; $Z = 17$.

8. Cr, chromium
   [2.11, 2.15]

The ion has a 2+ charge, i.e., two electrons, each of which has a negative charge, have been removed. The neutral atom should have 24 electrons; $Z = 24$.

9. N, nitrogen
   [2.14]

Each superscript indicates the number of electrons in the subshell. The 1s subshell has 2 electrons, the 2s subshell has 2 electrons, and the 2p subshell has 3 electrons. A total of 7 electrons indicates an atom with $Z = 7$, nitrogen. Check the periodic table in the back of the study guide.

10. Ni, nickel
    [2.14]

The total number of electrons in the atom is 28; $Z = 28$. Observe the position of nickel in the periodic table. The order of filling the atomic orbitals is shown up to $Z = 28$, and the corresponding spectroscopic notation is given to the left of the table.

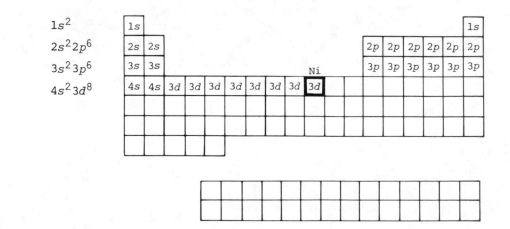

11. Kr, krypton [2.14]

The sum of the superscripts of the electronic configuration is 36; $Z = 36$.

12. Na, sodium [2.14, 2.15, 2.16]

$Z = 10$

13. Cu, copper [2.14, 2.15, 2.16]

$Z = 29$

14. C, carbon [2.15]

Since $n$, the principal quantum number, is 2, the element must be in the second row of the periodic table. Since $l = 1$, the element is one in which the $p$ subshell is being filled. The convention used in your text suggests that the $p$ orbitals are filled in the order given in Table 2.3 of the study guide.

TABLE 2.3   Order of Filling the $p$ Orbitals

| | | | Element | |
|---|---|---|---|---|
| Order | $m_l$ | $m_s$ | $n = 2, \; l = 1$ | $n = 3, \; l = 1$ |
| 1 | +1 | $+\frac{1}{2}$ | B | Al |
| 2 | 0 | $+\frac{1}{2}$ | C | Si |
| 3 | -1 | $+\frac{1}{2}$ | N | P |
| 4 | +1 | $-\frac{1}{2}$ | O | S |
| 5 | 0 | $-\frac{1}{2}$ | F | Cl |
| 6 | -1 | $-\frac{1}{2}$ | Ne | At |

15. Cl, chlorine

Elements in which the last electron is added to the $n = 3$ shell are indicated by the shaded area of the periodic table.

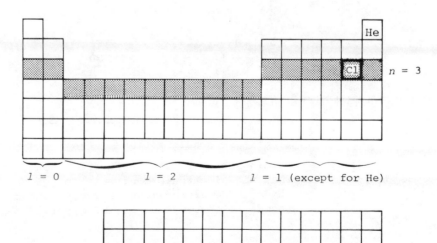

16. Pd, palladium      Elements in which the last electron is added to the
    [2.13]             $n = 4$ shell are indicated by the shaded area of the
                       periodic table. The convention used in your text sug-
                       gests that the $d$ orbitals ($l = 2$) are filled in the or-
                       der given in Table 2.4 of the study guide.

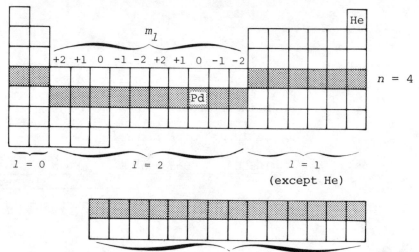

TABLE 2.4    Order of Filling the $d$ Orbitals

| | | | Element | |
| Order | $m_l$ | $m_s$ | $n = 4$, $l = 2$ | $n = 3$, $l = 2$ |
|---|---|---|---|---|
| 1 | +2 | $+\frac{1}{2}$ | Y | Sc |
| 2 | +1 | $+\frac{1}{2}$ | Zr | Ti |
| 3 | 0 | $+\frac{1}{2}$ | Nb | V |
| 4 | -1 | $+\frac{1}{2}$ | Mo | Cr |
| 5 | -2 | $+\frac{1}{2}$ | Tc | Mn |
| 6 | +2 | $-\frac{1}{2}$ | Ru | Fe |
| 7 | +1 | $-\frac{1}{2}$ | Rh | Co |
| 8 | 0 | $-\frac{1}{2}$ | Pd | Ni |
| 9 | -1 | $-\frac{1}{2}$ | Ag | Cu |
| 10 | -2 | $-\frac{1}{2}$ | Cd | Zn |

V. Electronic configurations

1. Ca: $1s^2 2s^2 2p^6-$      Reading the periodic table from left to right, we note
   $3s^2 3p^6 4s^2$           that the electronic configuration of Ca is $1s^2 2s^2 2p^6 3s^2-$
                              $3p^6 4s^2$. The arrangement of the periodic table corre-
                              sponds to the aufbau filling order.

$1s^2$

$2s^2 2p^6$

$3s^2 3p^6$

$4s^2$

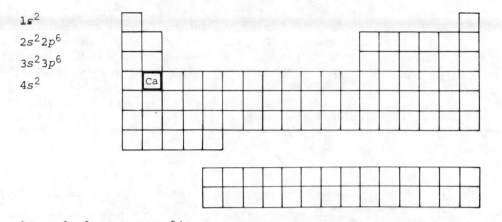

2. $Zn^{2+}$: $1s^2 2s^2-$
   $2p^6 3s^2 3p^6 3d^{10}$

For $Zn^{2+}$ write the electronic configuration of elemental zinc first. Always write a configuration in order of increasing principal quantum number; begin with the $s$ subshell, then write the $p$ subshell, the $d$ subshell, and finally the $f$ subshell of that shell. The configuration of Zn is $1s^2 2s^2 2p^6 3s^2 3p^6 3d^{10} 4s^2$.

$1s^2$

$2s^2 2p^6$

$3s^2 3p^6$

$4s^2 3d^{10}$

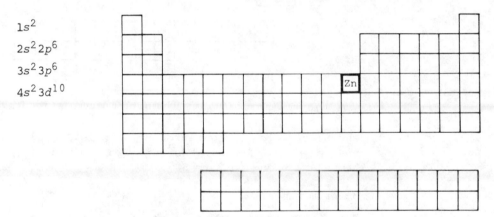

Form the ion by removing electrons from the outermost orbital. These are the ones on the extreme right in the electronic configuration. The electronic configuration of $Zn^{2+}$ is $1s^2 2s^2 2p^6 3s^2 3p^6 3d^{10}$. Notice that the aufbau order of filling orbitals is not the same as the order of removing electrons for ionization. Remember that the aufbau method is only a convention. Electron addition is but a small part of the total change: the elements are changing, protons and neutrons are being added to the nucleus, and $Z$ is increasing with each addition of a proton. In the formation of an ion, only electrons are added to or subtracted from the atom, the element remains the same, the nucleus is not changed, and $Z$ is not changed. *The order or adding or subtracting electrons to form ions is not and should not be expected to be the same as the aufbau order.* (See Section 2.14 of your text.)

3. $Cs^+$: $1s^2 2s^2$-
$2p^6 3s^2 3p^6 3d^{10}$-
$4s^2 4p^6 4d^{10}$-
$5s^2 5p^6$

The configuration of Cs is $1s^2 2s^2 2p^6 3s^2 3p^6 3d^{10} 4s^2 4p^6$-$4d^{10} 5s^2 5p^6 6s^1$ and that of the $Cs^+$ is $1s^2 2s^2 2p^6 3s^2 3p^6$-$3d^{10} 4s^2 4p^6 4d^{10} 5s^2 5p^6$.

$1s^2$

$2s^2 2p^6$

$3s^2 3p^6$

$4s^2 3d^{10} 4p^6$

$5s^2 4d^{10} 5p^6$

$6s^1$

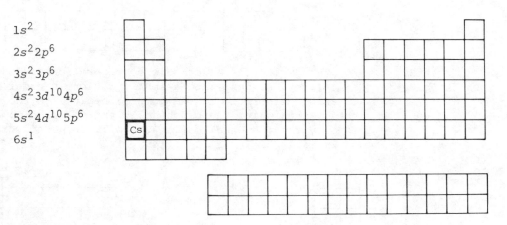

4. Kr: $1s^2 2s^2$-
$2p^6 3s^2 3p^6 3d^{10}$-
$4s^2 4p^6$

5. $S^{2-}$: $1s^2 2s^2$-
$2p^6 3s^2 3p^6$

The electronic configuration of S is $1s^2 2s^2 2p^6 3s^2 3p^4$ and that of $S^{2-}$ is $1s^2 2s^2 2p^6 3s^2 3p^6$. Two additional $3p$ electrons are added to form $S^{2-}$.

VI. Quantum numbers

1. O: $n = 2$,
   $l = 1$,
   $m_l = +1$,
   $m_s = -\frac{1}{2}$

The last electron added is the fourth one in the $2p$ subshell. Therefore, $n = 2$ and $l = 1$. The quantum numbers $m_l$ and $m_s$ by the convention used in your text are +1 and $-\frac{1}{2}$, respectively (see Table 2.3 of the study guide). The set of quantum numbers that corresponds to the last electron added is

$$n = 2, \ l = 1, \ m_l = +1, \ m_s = -\tfrac{1}{2}$$

The periodic table in Figure 2.1 of the study guide may be helpful in the determination of the quantum numbers of the last electron added in the building of elements when the aufbau method is used.

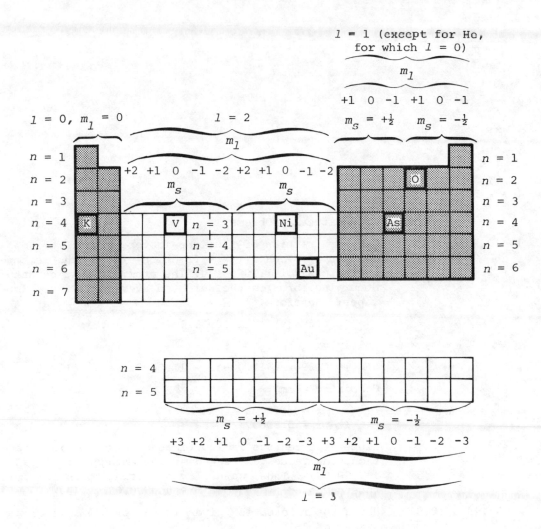

FIGURE 2.1  Periodic Table

2. K: $n = 4$,
   $l = 1$,
   $m_l = 0$
   $m_s = +\frac{1}{2}$

3. V: $n = 3$,
   $l = 2$,
   $m_l = 0$,
   $m_s = +\frac{1}{2}$

4. As: $n = 4$,
   $l = 1$,
   $m_l = -1$,
   $m_s = +\frac{1}{2}$

5. Ni: $n = 3$,
   $l = 2$,
   $m_l = 0$,
   $m_s = -\frac{1}{2}$

6. Au: $n = 5$,
   $l = 2$,
   $m_l = -1$,
   $m_s = -\frac{1}{2}$

## VII. Mathematical problems

1. 19.903% $^{10}$B
   80.097% $^{11}$B
   [2.18]

   Set $x$ = fraction of $^{10}$B, the isotope with a mass of 10.01294 u. Then $1 - x$ = fraction of $^{11}$B, the isotope with a mass of 11.00931 u. The atomic weight of naturally occurring boron is the weighted average of these isotopes; therefore,

   $$10.01294(x) + 11.00931(1 - x) = 10.811$$

   Multiplying terms,

   $$10.01294x + 11.00931 - 11.00931x = 10.811$$

   Combining terms,

   $$x = 0.19903$$
   $$1 - x = 0.80097$$

   Since a percentage is 100 times a corresponding fraction, the values expressed as percentages are 19.903% and 80.097%. Thus, naturally occurring boron is a mixture composed of 19.903% $^{10}$B and 80.097% $^{11}$B.

2. 35.49 u [2.9]

   Using the information given in the problem, we obtain an answer of 35.49 u:

   $$0.7553(35.00 \text{ u}) + 0.2447(37.00 \text{ u}) = 35.49 \text{ u}$$

   This answer is very close to the actual atomic weight, 35.45 u.

3. $3 \times 10^2$ Mhz
   [2.10]

   Since frequency is inversely proportional to wavelength, $\lambda$, and directly proportional to the velocity of light, $c$:

   $$\nu = c/\lambda$$

   and find

   $$\nu = (2.9979 \times 10^{10} \text{ cm sec}^{-1})/100 \text{ cm}$$
   $$\nu = (3.0 \times 10^{10} \text{ cm sec}^{-1})/(1 \times 10^1 \text{ cm})$$

Notice that units are included with all numbers and that the number of significant figures is adjusted according to the procedure outlined in Chapter 1 of your text. The frequency can be expressed in reciprocal seconds, since the unit of length, the centimeter, can be canceled:

$$\nu = (3.0 \times 10^{10} \text{ cm sec}^{-1})/(1 \times 10^2 \text{ cm})$$
$$\nu = 3 \times 10^8 \text{ sec}^{-1}$$

Notice that the frequency is expressed in one significant figure, because the calculation involves a value, 100 cm, that has only one significant figure. Since a reciprocal second is a hertz and $10^6$ hertz is a megahertz, the frequency can be expressed as $3 \times 10^2$ Mhz:

$$\nu = (3 \times 10^8 \text{ hz})(1 \text{ Mhz}/10^6 \text{ hz})$$
$$\nu = 3 \times 10^2 \text{ Mhz}$$

4. 1.0 nm X ray
   [2.10]

Substituting values into the equation

$$E = \frac{hc}{\lambda}$$

we obtain

$$E = \frac{(6.63 \times 10^{-34} \text{ J sec})(3.0 \times 10^{10} \text{ cm sec}^{-1})}{1.0 \times 10^{-7} \text{ cm}}$$

$$= 2.0 \times 10^{-16} \text{ J for the X ray.}$$

The energy of the microwave is calculated from the equation

$$E = h\nu$$

Substituting,

$$E = (6.63 \times 10^{-34} \text{ J sec})(3.0 \times 10^{10} \text{ sec}^{-1})$$
$$= 2.0 \times 10^{-23} \text{ J}$$

Thus, the X ray has the higher energy.

## SELF-TEST

Complete the test in 45 minutes:

I. Fill in the blank space with the word that most appropriately completes the statement or answers the question:

1. _____ equals Planck's constant, $h$, times the frequency, $\nu$, of the electromagnetic radiation.

2. The fundamental subatomic particles are the _____, _____, and _____.

3. The fundamental subatomic particle that has no charge is the _____.

4. Atoms with the same atomic number but different mass numbers are called _____.

5. The maximum number of electrons that can exist in the $M$ shell of any atom is _____.

6. A material is said to exhibit what type of magnetic behavior if all electrons are paired? _____

7. The electronic configuration of lithium is _____ and the value of $Z$ of the element is _____.

II. Do the following:

1. Define the atomic mass unit.

2. Write the quantum numbers of the last electron that is added to iridium, Ir, according to the aufbau method. Use a periodic table.

3. Write the electronic configuration of Ir and that of $Ir^{2+}$.

4. Determine the atomic weight of lithium. Lithium is a mixture of two naturally occurring isotopes; 7.40% of the mixture is $^6Li$, an isotope that has a mass of 6.0169 u, and 92.60% of the mixture is $^7Li$, an isotope that has a mass of 7.0182 u.

III. Answer each of the following:

_____

1. The differentiating electron of a sulfur atom has which of the following sets of quantum numbers?
a. $n = 3$, $l = 1$, $m_l = +1$, $m_s = +\frac{1}{2}$
b. $n = 3$, $l = 1$, $m_l = -1$, $m_s = +\frac{1}{2}$
c. $n = 3$, $l = 1$, $m_l = +1$, $m_s = -\frac{1}{2}$
d. $n = 3$, $l = 1$, $m_l = -1$, $m_s = -\frac{1}{2}$

_____

2. Which of the following sets of quantum numbers represents an impossible arrangement?
a. $n = 2$, $l = 0$, $m_l = 0$, $m_s = -\frac{1}{2}$
b. $n = 7$, $l = 4$, $m_l = -1$, $m_s = +\frac{1}{2}$
c. $n = 3$, $l = 1$, $m_l = +1$, $m_s = -\frac{1}{2}$
d. $n = 3$, $l = -1$, $m_l = -1$, $m_s = -\frac{1}{2}$

3. The success of J. J. Thomson's experiment to measure
   the charge-to-mass ratio of the electron depended on
   a. the behavior of an electron in an electrical field.
   b. the behavior of an electron in a magnetic field.
   c. the ability to detect the exact point of impact
      of a stream of electrons with matter.
   d. all of the above

4. The discovery and characterization of the neutron
   occurred significantly later than that of either the
   proton or the electron. This can be attributed to
   the fact that
   a. the mass of the neutron is so similar to that of
      the proton that the two particles are essentially
      indistinguishable.
   b. neutrons are buried deep within the nucleus and
      are not easily accessible.
   c. neutrons are uncharged particles.
   d. neutrons have a very short mean lifetime.

5. In which of the following sets do the nuclides have
   an isotopic relationship to one another?
   a. nuclide I (7 protons, 6 neutrons)
      nuclide II (6 protons, 7 neutrons)
   b. nuclide I (7 protons, 7 electrons)
      nuclide II (6 protons, 6 electrons)
   c. nuclide I (7 protons, 6 neutrons)
      nuclide II (7 protons, 7 neutrons)
   d. nuclide I (7 protons, 6 neutrons)
      nuclide II (6 protons, 6 neutrons)

6. Which two sets in problem 5 contain nuclides with
   the same charge?

7. The wave mechanical approach to atomic structure per-
   mits the calculation of
   a. a volume about the nucleus in which an electron
      of specified energy will most probably be found.
   b. the most probable radius of an orbit that an
      electron of specified energy will follow.
   c. the most probable position of an electron of
      specified energy at a given time, $t$.
   d. the most probable spin value that will be asso-
      ciated with an electron of specified energy.

8. As electrons move from a ground state to an excited
   state,
   a. an emission spectrum results.
   b. energy is absorbed.
   c. heat is liberated.
   d. light is emitted.

_____ 9. The intensity of a spectral line observed in an atomic emission spectrum can be directly related to
a. the difference in energy of the energy levels involved in the electron transition that gives rise to the line.
b. the number of electrons undergoing the transition that gives rise to the line.
c. the number of energy levels involved in the transition that gives rise to the line.
d. the speed with which an electron undergoes a transition from one energy level to another.

_____ 10. The wavelength at which a spectral line is observed in an atomic emission spectrum is inversely related to
a. the difference in energy of the energy levels involved in the electron transition that gives rise to the line.
b. the number of electrons undergoing the transition that gives rise to the line.
c. the number of energy levels involved in the transition that gives rise to the line.
d. the speed with which an electron undergoes a transition from one energy level to another.

_____ 11. The quantum number that designates the spin of an electron is
a. $n$        b. $l$        c. $m_l$        d. $m_s$

_____ 12. How do chemists know that the atomic weight of the $^{12}C$ nuclide is exactly 12.000?
a. The $^{12}C$ nuclide weighs exactly 12 times as much as hydrogen.
b. $6.02 \times 10^{12}$ atoms of $^{12}C$ weigh exactly 12 grams.
c. The $^{12}C$ nuclide is composed of 6 protons and 6 neutrons, each weighing 1 unit.
d. Chemists define it so.

_____ 13. Which of the following is the most paramagnetic?
a. $K^+$        b. $Zn^{2+}$        c. $Cu^{2+}$        d. $Fe^{3+}$

_____ 14. Scandium
a. is a transition element.
b. has an atomic number of 45.
c. has 21 electrons in the $Sc^{2+}$ ion.
d. has a completely filled $n = 3$ subshell.

_____ 15. Which of the following is diamagnetic?
a. K        b. $Na^+$        c. $Co^{2+}$        d. P

# Chemical Bonding

OBJECTIVES

I. You should be able to demonstrate your knowledge of the following terms by defining them, describing them, or giving specific examples of them:

anion [3.4]
atomic radius [3.1]
bond energy [3.9]
cation [3.4]
covalent bond [3.6]
dipole moment [3.8]
electron affinity [3.3]
electronegativity [3.9]
electrovalence number [3.4]
formal charge [3.7]
ionic compound [3.4]
ionic radius [3.6]
ionization energy [3.2]

                    isoelectronic [3.4]
                    lanthanide contraction [3.1]
                    lattice energy [3.5]
                    Lewis structure [3.7, 3.8, 3.9]
                    nomenclature [3.13]
                    oxidation number [3.12]

    II. Given the names or chemical formulas of common compounds,
        you should be able to write the corresponding chemical
        formulas or names.

   III. Using the periodic table of the elements, you should be
        able to predict relative sizes of atoms and ions and
        relative magnitudes of ionization energies and electro-
        negativities.

    IV. You should be able to predict the common oxidation num-
        bers of elements and to determine oxidation numbers of
        these elements in polyatomic molecules and ions.

     V. You should be able to draw Lewis structures of molecules
        and ions and to determine the formal charge of each atom
        in such structures.

    VI. You should be able to predict which bond in a given set
        of compounds has the most covalent character.

   VII. You should be able to calculate the partial ionic charac-
        ter of a bond from dipole moment data and bond distance
        data. You should also be able to determine the polarity
        of a bond from the electronegativities of the elements
        that are bonded.

## EXERCISES

   I. Write the formula of each of the following compounds in
      the space provided:

   _____     1. potassium sulfide
   _____     2. sodium chloride
   _____     3. magnesium sulfide
   _____     4. calcium oxide
   _____     5. barium sulfide
   _____     6. strontium iodide
   _____     7. beryllium fluoride
   _____     8. sodium sulfide
   _____     9. ammonium iodide
   _____    10. lithium hydroxide
   _____    11. cesium iodide

_____  12. sodium cyanide
_____  13. potassium dichromate
_____  14. sodium bromide
_____  15. calcium carbonate
_____  16. potassium phosphate
_____  17. sodium permanganate
_____  18. hydrobromic acid
_____  19. hydrobromous acid
_____  20. bromous acid
_____  21. bromic acid
_____  22. perbromic acid
_____  23. potassium perbromate
_____  24. barium sulfate
_____  25. sodium nitride
_____  26. carbon disulfide
_____  27. diiodine pentoxide
_____  28. nitrogen oxide
_____  29. phosphorus trifluoride
_____  30. tetraarsenic tetrasulfide
_____  31. iodine trifluoride
_____  32. dichlorine oxide
_____  33. carbon dioxide
_____  34. sulfur dioxide
_____  35. dinitrogen pentoxide
_____  36. tin(II) bromide
_____  37. tin(IV) bromide
_____  38. mercury(I) sulfide
_____  39. gold(III) chloride
_____  40. diantimony trioxide
_____  41. ferrous sulfate
_____  42. ferric sulfate
_____  43. copper(II) dihydrogen phosphate
_____  44. titanium(IV) chloride
_____  45. tin(II) oxide
_____  46. stannic oxide
_____  47. calcium hydride
_____  48. sodium peroxide
_____  49. calcium phosphide
_____  50. sodium monohydrogen phosphite
_____  51. xenon hexafluoride
_____  52. sodium bisulfate
_____  53. barium arsenate
_____  54. potassium arsenite

II. Write the name of each of the following compounds in
    the space provided:

_____  1. $NF_3$
_____  2. NaCl

_____    3. NaClO
_____    4. NaClO$_2$
_____    5. NaClO$_3$
_____    6. NaClO$_4$
_____    7. CCl$_4$
_____    8. CaCrO$_4$
_____    9. ZnHPO$_4$
_____   10. Sn$_3$(PO$_4$)$_2$
_____   11. N$_2$F$_4$
_____   12. Cu(NO$_3$)$_2$
_____   13. Ba$_3$N$_2$
_____   14. TiCl$_2$
_____   15. NH$_3$
_____   16. MnO$_2$
_____   17. N$_2$O$_5$
_____   18. BaCr$_2$O$_7$

III. Answer the following:

1. Which has the larger radius?

_____  a. Mg or Si            _____  d. Ti or Cr
_____  b. Li or Cs            _____  e. Mg or Mg$^{2+}$
_____  c. Ni or Zn            _____  f. Cl or Cl$^-$

2. Which has the larger electronegativity?

_____  a. K or Br            _____  b. O or Te

3. Which has the larger first ionization energy?

_____  a. Ba or Bi           _____  d. Br or Kr
_____  b. Al or Tl           _____  e. P or O
_____  c. C or O

4. Which has the larger first electron affinity?

_____  a. B or F             _____  b. Cl or I

5. Using the data in Figure 3.1 of the study guide,
   predict the degree of polarity of a bond between
   each of the following pairs of atoms. Arrange them
   in order of increasing polarity.

_____  a. C and I            _____  d. S and Cl
_____  b. Al and Cl          _____  e. Cs and S
_____  c. K and Cl           _____  f. Ca and O

6. Using the concept of anion deformation, predict
   which one of the two compounds in each of the following pairs has the bond with the greater amount of
   covalent character?

_____  a. SlCl$_3$ or BiCl$_3$       _____  e. ZnS or CdS
_____  b. CrCl$_2$ or CrCl$_3$       _____  f. Tl$_2$O or Tl$_2$O$_3$
_____  c. HgCl$_2$ or HgI$_2$        _____  g. MgO or CaO
_____  d. BeO or BeS                 _____  h. SO$_2$ or SeO$_2$

| 1<br>H<br>2.1 | | | | | | | | | | | | | | | | | 2<br>He<br>— |
|---|---|---|---|---|---|---|---|---|---|---|---|---|---|---|---|---|---|
| 3<br>Li<br>1.0 | 4<br>Be<br>1.5 | | | | | | | | | | | 5<br>B<br>2.0 | 6<br>C<br>2.5 | 7<br>N<br>3.0 | 8<br>O<br>3.5 | 9<br>F<br>4.0 | 10<br>Ne<br>— |
| 11<br>Na<br>0.9 | 12<br>Mg<br>1.2 | | | | | | | | | | | 13<br>Al<br>1.5 | 14<br>Si<br>1.8 | 15<br>P<br>2.1 | 16<br>S<br>2.5 | 17<br>Cl<br>3.0 | 18<br>Ar<br>— |
| 19<br>K<br>0.8 | 20<br>Ca<br>1.0 | 21<br>Sc<br>1.3 | 22<br>Ti<br>1.5 | 23<br>V<br>1.6 | 24<br>Cr<br>1.6 | 25<br>Mn<br>1.5 | 26<br>Fe<br>1.8 | 27<br>Co<br>1.8 | 28<br>Ni<br>1.8 | 29<br>Cu<br>1.9 | 30<br>Zn<br>1.6 | 31<br>Ga<br>1.6 | 32<br>Ge<br>1.8 | 33<br>As<br>2.0 | 34<br>Se<br>2.4 | 35<br>Br<br>2.8 | 36<br>Kr<br>— |
| 37<br>Rb<br>0.8 | 38<br>Sr<br>1.0 | 39<br>Y<br>1.2 | 40<br>Zr<br>1.4 | 41<br>Nb<br>1.6 | 42<br>Mo<br>1.8 | 43<br>Tc<br>1.9 | 44<br>Ru<br>2.2 | 45<br>Rh<br>2.2 | 46<br>Pd<br>2.2 | 47<br>Ag<br>1.9 | 48<br>Cd<br>1.7 | 49<br>In<br>1.7 | 50<br>Sn<br>1.8 | 51<br>Sb<br>1.9 | 52<br>Te<br>2.1 | 53<br>I<br>2.5 | 54<br>Xe<br>— |
| 55<br>Cs<br>0.7 | 56<br>Ba<br>0.9 | 57–71<br>La–Lu<br>1.1–1.2 | 72<br>Hf<br>1.3 | 73<br>Ta<br>1.5 | 74<br>W<br>1.7 | 75<br>Re<br>1.9 | 76<br>Os<br>2.2 | 77<br>Ir<br>2.2 | 78<br>Pt<br>2.2 | 79<br>Au<br>2.4 | 80<br>Hg<br>1.9 | 81<br>Tl<br>1.8 | 82<br>Pb<br>1.8 | 83<br>Bi<br>1.9 | 84<br>Po<br>2.0 | 85<br>At<br>2.2 | 86<br>Rn<br>— |
| 87<br>Fr<br>0.7 | 88<br>Ra<br>0.9 | 89–<br>Ac–<br>1.1–1.7 | | | | | | | | | | | | | | | |

FIGURE 3.1 Electronegativities of the Elements. (Based on Linus Pauling, *The Nature of the Chemical Bond,* Third Edition, © 1960 by Cornell University Press. Used with the permission of Cornell University Press.)

IV. Answer each of the following with *true* or *false*. If a statement is false, correct it.

_____  1. An atom of any group V A element has five valence electrons.

_____  2. The oxidation number of each of the alkali metal ions is 2+.

_____  3. The most reactive nonmetal on the periodic chart is the fluorine atom.

_____  4. The ionization energy of a noble gas is very high compared to that of any of the other elements in the same period.

_____  5. The first electron affinity of a halogen is low compared to that of other elements in the same period. That is, less energy is evolved when an electron is added to a halogen atom than when an electron is added to any other element in the same period.

_____  6. Monatomic anions are named by replacing the last part of the element's name with "ide."

_____   7. Ionization energy is the amount of energy required to add an electron to an isolated gaseous atom in its ground state.

_____   8. The element in the periodic table that has the highest electronegativity is fluorine.

_____   9. $Na^+$ and Ne have the same electronic structure.

_____   10. Purely ionic bonds are formed by two atoms sharing electrons.

_____   11. The second ionization energy of sodium should be larger than the second ionization energy of magnesium.

_____   12. The more polar the bond in a diatomic molecule the lower the dipole moment of that molecule.

V. Draw the most probable Lewis structure of each of the following and include the formal charges. The formulas of the more complex, multi-element molecules are written to indicate the general atomic arrangement of the molecule. Thus, $N_2F_2$ is written as FNNF to show that the two nitrogen atoms are joined by at least one chemical bond and that each nitrogen atom forms at least one bond with a single fluorine atom.

_____   1. $O_2$

_____   2. $O_3$

_____   3. $CH_4$

_____   4. $N_3^-$

_____   5. $OF_2$

_____   6. $H_2PO_2^-$

_____   7. FNNF

_____   8. $C_2^{2-}$

_____   9. $BF_4^-$

_____   10. $NH_4^+$

_____   11. $NO_3^-$

_____   12. $SO_4^{2-}$

_____   13. $N_2$

_____   14. $ClO^-$

_____   15. NNO

_____   16. $O_2SF_2$

_____   17. $H_2NNH_2$

_____   18. $SnCl_4$

_____   19. $OH^-$

_____   20. $O_2NONO_2$

_____   21. $ONNO^{2-}$

_____   22. $H_3COCH_3$

_____   23. $H_2CO$

_____   24. $XeO_3$

VI. What is the oxidation number of:

_____ 1. B in $BF_4^-$

_____ 2. N in $NH_4^+$

_____ 3. N in $NO_3^-$

_____ 4. P in $H_2PO_2^-$

_____ 5. P in $PO_4^{3-}$

_____ 6. Cl in $ClO_4^-$

_____ 7. Sn in $SnCl_4$

_____ 8. Cr in $Cr_2O_7^{2-}$

_____ 9. Cr in $CrO_4^{2-}$

_____ 10. Mn in $MnO_2$

_____ 11. Mn in $MnO_4^-$

_____ 12. Cl in $Cl_2$

## ANSWERS TO EXERCISES

I. Formulas of compounds

1. $K_2S$
2. NaCl

Each element of Group I A (see Figure 3.2 of the study guide for this and subsequent exercises) has one electron in its valence shell and an oxidation number of 1+ in its compounds.

3. MgS
4. CaO

Each element of Group II A has two electrons in its valence shell and an oxidation number of 2+ in its compounds.

5. BaS

Each element of group VI A has six electrons in its valence shell and usually an oxidation number of 2– in its compounds.

| I A | II A | | | | | | | | | | | | III A | IV A | V A | VI A | VII A | O |
|---|---|---|---|---|---|---|---|---|---|---|---|---|---|---|---|---|---|---|
| H | | | | | | | | | | | | | | | | | | He |
| Li | Be | Transition Elements | | | | | | | | | | | B | C | N | O | F | Ne |
| Na | Mg | | | | | | | | | | | | Al | Si | P | S | Cl | Ar |
| K | Ca | Sc | Ti | V | Cr | Mn | Fe | Co | Ni | Cu | Zn | | Ga | Ge | As | Se | Br | Kr |
| Rb | Sr | Y | Zr | Nb | Mo | Tc | Ru | Rh | Pd | Ag | Cd | | In | Sn | Sb | Te | I | Xe |
| Cs | Ba | La* | Hf | Ta | W | Re | Os | Ir | Pt | Au | Hg | | Tl | Pb | Bi | Po | At | Rn |
| Fr | Ra | Ac** | | | | | | | | | | | | | | | | |

Metals / Nonmetals

Inner-Transition Elements

| * | Ce | Pr | Nd | Pm | Sm | Eu | Gd | Tb | Dy | Ho | Er | Tm | Yb | Lu |
|---|---|---|---|---|---|---|---|---|---|---|---|---|---|---|
| ** | Th | Pa | U | Np | Pu | Am | Cm | Bk | Cf | Es | Fm | Md | No | Lr |

FIGURE 3.2  Periodic Table of Elements. (Nonmetals are within darkly shaded area; metals within lightly shaded area commonly display more than one oxidation state; metals within unshaded area usually display only one oxidation state, +1, +2, or +3.)

6. $SrI_2$

7. $BeF_2$

Each element of Group VII A has seven electrons in its valence shell and usually an oxidation number of 1- in its compounds.

8. $Na_2S$

9. $NH_4I$

An ammonium ion has a charge of 1+.

10. $LiOH$

A hydroxide ion has a charge of 1-.

11. $CsI$

12. $NaCN$

A cyanide ion has a charge of 1-.

13. $K_2Cr_2O_7$

A dichromate ion has a charge of 2-.

14. $NaBr$

15. $CaCO_3$

16. $K_3PO_4$

17. $NaMnO_4$

A permanganate ion has a charge of 1-.

18. $HBr$

19. $HBrO$

20. $HBrO_2$

21. $HBrO_3$

22. $HBrO_4$

These are the formulas of oxybromine acids. Formulas for oxyacids of other elements are given in Table 3.1 of the study guide.

23. $KBrO_4$

The perbromate ion has a charge of 1-.

24. $BaSO_4$

The sulfate ion has a charge of 2-.

25. $Na_3N$

In this compound the oxidation number of nitrogen is 3-.

26. $CS_2$

27. $I_2O_5$

Prefixes are used in the names of covalent compounds to indicate the number of atoms of each element in a molecule of that compound. In general, this type of nomenclature is used for any combination of nonmetals. The darkly shaded area of the periodic table in Figure 3.2 of the study guide indicates the elements that are nonmetals.

28. $NO$

29. $PF_3$

TABLE 3.1   Some Common Oxyacids

| Name | Formula | Name | Formula |
|---|---|---|---|
| sulfurous acid | $H_2SO_3$ | nitrous acid | $HNO_2$ |
| sulfuric acid | $H_2SO_4$ | nitric acid | $HNO_3$ |
| hypophosphorous acid | $H_3PO_2$ | hypoiodous acid | $HIO$ |
| phosphorous acid | $H_3PO_3$ | iodic acid | $HIO_3$ |
| phosphoric acid | $H_3PO_4$ | periodic acid | $HIO_4$ |

30. $As_4S_4$

31. $IF_3$

32. $Cl_2O$

33. $CO_2$

34. $SO_2$

35. $N_2O_5$

36. $SnBr_2$
37. $SnBr_4$     Metals that commonly display more than one oxidation
state are indicated by the lightly shaded area in the
periodic table of Figure 3.2 of the study guide. The
oxidation number of such an element in a compound can be
indicated in the name of that compound by roman numerals
in parentheses after the English name of that element.
Thus, $SnBr_2$ is named tin(II) bromide, and $SnBr_2$ is named
tin(IV) bromide.

38. $Hg_2S$

39. $AuCl_3$

40. $Sb_2O_3$

41. $FeSO_4$

42. $Fe_2(SO_4)_3$

43. $Cu(H_2PO_4)_3$

44. $TiCl_4$

45. SnO
46. $SnO_2$     The suffixes -ous and -ic are often used to indicate
the lower and the higher oxidation state of an ele-
ment, respectively. Names such as this should be memo-
rized.

47. $CaH_2$     Hydrogen sometimes has an oxidation number of 1- in its
compounds.

48. NaO     Oxygen sometimes has an oxidation number of 1- in its
compounds.

49. $Ca_3P_2$

50. $Na_2HPO_3$

51. $XeF_6$

52. $NaHSO_4$

53. $Ba_3(AsO_4)_2$

54. $K_3AsO_3$

II. Names of compounds

The preferred names are given for the compounds.

1. nitrogen trifluoride
2. sodium chloride
3. sodium hypochlorite
4. sodium chlorite    The names of anions of other common oxyacids are given
5. sodium chlorate    in Table 3.2 of the study guide.
6. sodium perchlorate
7. carbon tetrachloride
8. calcium chromate
9. zinc(II) hydro-genphosphate
10. tin(II) phosphate
11. dinitrogen tetrafluoride
12. copper(II) nitrate or cupric nitrate
13. barium nitride
14. titanium(II) chloride
15. ammonia    This common name is preferred.
16. manganese(IV) oxide
17. dinitrogen pentoxide
18. barium dichromate

TABLE 3.2   Anions of Some Common Oxyacids

| Name | Formula | Name | Formula |
|------|---------|------|---------|
| sulfate ion | $SO_4^{2-}$ | nitrite ion | $NO_2^-$ |
| sulfite ion | $SO_3^{2-}$ | nitrate ion | $NO_3^-$ |
| hypophosphite ion | $PO_2^{3-}$ | hypoiodite ion | $IO^-$ |
| phosphite ion | $PO_3^{3-}$ | iodate ion | $IO_3^-$ |
| phosphate ion | $PO_4^{3-}$ | periodate ion | $IO_4^-$ |

III. Atomic properties as related to chemical bonding

1. [3.1]  a. Mg    There is generally a decrease in atomic radius across a period from left to right.

   [3.1]  b. Cs    There is generally an increase in atomic radius in a group from top to bottom.

   [3.1]  c. Zn    In the transition series inner electrons of the atoms screen the outer electrons from the nuclear charge. Toward the end of the series the atomic radius actually increases with increasing atomic number.

   [3.1]  d. Ti    The screening effect of inner electrons is not so pronounced in the beginning of a transition series, and the atomic radius decreases with increasing atomic number.

   [3.6]  e. Mg    The magnesium ion, $Mg^{2+}$, has the same nuclear charge as the magnesium atom but two fewer electrons. The radius of the magnesium ion is therefore less than that of the magnesium atom.

   [3.6]  f. $Cl^-$    The chloride ion, $Cl^-$, has the same nuclear charge as the chlorine atom but one additional electron. The radius of the chloride ion is therefore larger than that of the chlorine atom.

2. [3.9]  a. Br    There is generally an increase in electronegativity across a period from left to right.

   [3.9]  b. O     There is generally a decrease in electronegativity in a group from top to bottom.

3. [3.2]  a. Bi    There is generally an increase in ionization energy across a period from left to right.

          b. Al    There is generally a decrease in ionization energy from top to bottom.

          c. O     See the explanation given for part (a) of this question.

          d. Kr    Krypton is a noble gas and has a full outer shell of electrons. A large amount of energy would be required to remove an electron and form a less stable ion.

          e. O     See the explanations given for parts (a) and (b) of this question.

4. [3.3]  a. F     Electron affinity generally increases across a period from left to right; however, there are significant exceptions. See Figure 3.3 of the study guide.

          b. Cl    Electron affinity generally decreases in a group from top to bottom; however, there are exceptions. See Figure 3.3 of the study guide.

| H −72 | | | | | | | He +54 |
|---|---|---|---|---|---|---|---|
| Li −57 | Be +66 | B −15 | C −121 | N +31 | O −142 (+702)* | F −333 | Ne +99 |
| Na −21 | Mg +67 | Al −26 | Si −135 | P −60 | S −200 (+332)* | Cl −348 | |
| | | | | | | Br −324 | |
| | | | | | | I −295 | |

FIGURE 3.3  Electron Affinities (kJ/mol). (Values in parentheses pertain to the *total* energy effect for the addition of *two* electrons.)

5. [3.11]
   a, d, b,
   e, c, f

The descriptive terms that follow are the authors personal denotation of the trend from covalent to ionic bonds. There are no specific terms to indicate the degree of polarity that a bond possesses, but the trend from a to f should be reflected.

a. Covalent, nonpolar: The electronegativity difference = 2.5 - 2.5 = 0, and the bond is predicted to be covalent and nonpolar.

b. Covalent, moderately polar: The electronegativity difference = 3.0 - 1.5 = 1.5, and the bond is predicted to be covalent and moderately polar.

c. Ionic: The electronegativity difference = 3.0 - 0.8 = 2.2, and the bond is predicted to be ionic.

d. Covalent, somewhat polar: The electronegativity difference = 3.0 - 2.5 = 0.5, and the bond is predicted to be covalent and somewhat polar.

e. Covalent, highly polar: The electronegativity difference = 2.5 - 0.7 = 1.8, and the bond is predicted to be covalent and highly polar.

f. Very ionic: The electronegativity difference = 3.5 - 1.0 = 2.5, and the bond is predicted to be ionic.

6. [3.8]

The circles in this section are drawn to scale in order to show relative ionic sizes. Remember that only anion distortion is considered and that the more distorted anion is forming a more covalent bond.

a. $SbCl_3$

Since $Sb^{3+}$ is smaller than $Bi^{3+}$, $Sb^{3+}$ has a larger charge density, concentration of charge, than $Bi^{3+}$. As a result, $Sb^{3+}$ distorts, i.e., attracts, the electron cloud of the chlorine ion more than the $Bi^{3+}$ does.

$Sb^{3+}$     $Cl^-$                    $Bi^{3+}$     $Cl^-$

The larger the cation charge concentration the more distorted the anion and the more covalent the bond.

b. $CrCl_3$

$Cr^{3+}$ has a larger charge density than $Cr^{2+}$; therefore, $Cr^{3+}$ distorts the anion more than $Cr^{2+}$ does and thus forms the more covalent bond.

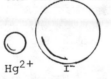   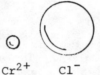

$Cr^{3+}$     $Cl^-$                    $Cr^{2+}$     $Cl^-$

Larger cation charge           Smaller cation charge
  concentration                    concentration
More distorted anion           Less distorted anion
More covalent                  Less covalent

c. $HgI_2$

The larger $I^-$ ion can be more easily distorted than the smaller $Cl^-$ ion.

$Hg^{2+}$     $I^-$                     $Hg^{2+}$     $Cl^-$

Larger anion                   Smaller anion
More distortion of anion       Less distortion of anion
More covalent                  Less covalent

d. BeS

The sulfide anion is larger than the oxide anion and is therefore more easily distorted than the oxide anion.

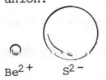

$Be^{2+}$     $S^{2-}$                  $Be^{2+}$     $O^{2-}$

Larger anion                   Smaller anion
More easily distorted          Less easily distorted
  anion                            anion
More covalent bond             Less covalent bond

e. ZnS          See the explanation given for part (a) of this problem.

f. $Tl_2O_3$        See the explanation given for part (b) of this problem.

g. MgO          See the explanation given for part (a) of this problem.

h. $SO_2$          See the explanation given for part (a) of this problem.

IV. Concepts of chemical bonding

1. True [3.8, 3.9]    Valence electrons are those electrons in the outermost shell of an element. All elements of group V A (N, P, As, Sb, and Bi) has an $ns^2np^3$ electronic configuration.

2. False [3.12]    The oxidation number of each of the alkali metal ions ($Li^+$, $Na^+$, $K^+$, $Rb^+$, $Cs^+$, $Fr^+$) is 1+.

3. True [3.11]    Also see Figure 3.13 in your text.

4. True [3.2]    A noble gas has a full outer shell, $ns^2np^6$; a very large amount of energy is required to ionize such an atom.

5. False [3.3]    The electron affinity is the energy evolved when an electron is added to an atom. More energy is evolved in the formation of a halide anion than for other elements in the same period.

6. True [3.13]    See Table 3.3 of the study guide for the names of anions of some common elements.

7. False [3.2, 3.3]    The first ionization energy is the amount of energy necessary to remove an electron from an isolated gaseous atom in its ground state, $X \rightarrow X^+ + e^-$. Electron affinity is the energy effect accompanying the process in which an electron is added to an isolated atom in its ground state, $X + e^- \rightarrow X^-$.

8. True [3.9]    The general trend of electronegativity is an increase from left to right across a period and a decrease from top to bottom within a group.

TABLE 3.6    Names of Anions of Some Common Elements

| Element Name | Symbol | Anion | Anion Name |
|---|---|---|---|
| fluorine | F | $F^-$ | fluoride |
| nitrogen | N | $N^{3-}$ | nitride |
| oxygen | O | $O^{2-}$ | oxide |
| carbon | C | $C^{4-}$ | carbide |
| sulfur | S | $S^{2-}$ | sulfide |
| bromine | Br | $Br^-$ | bromide |
| selenium | Se | $Se^{2-}$ | selenide |

9. True [3.4]    The term *isoelectronic* is used to indicate that species have the same electronic configuration. For example,

$$Na^+: 1s^2 2s^2 2p^6$$
$$Ne : 1s^2 2s^2 2p^6$$

are said to be isoelectronic.

10. False [3.4]    Covalent bonds are those bonds formed by the sharing of electrons. Ionic bonds are those bonds formed by the attraction of oppositely charged ions.

11. True [3.5]    Singly ionized sodium has a stable configuration, a noble gas structure, but magnesium only attains this stable configuration after being doubly ionized.

12. False [3.11]    The more polar a bond, the larger its charge separation, and therefore the larger its dipole moment.

V. Lewis structures [3.9]

Follow the steps outlined in Section 3.9 of the text:

1. Find the total number of valence electrons in the atoms forming the molecule.
2. Find the total number of electrons that are needed to fill the valence shell of all atoms in the molecule.
3. Compute the number of electrons that must be shared to obtain filled valence shells in all atoms (value from step 2 minus value from step 1).
4. Determine the number of bonds. Each pair of electron that need to be shared represents one bond (one-half the value determined in step 3).
5. Draw the structure with the proper number of bonds. First place a single bond between bonded atoms; if more bonds are needed, add multiple bonds. There may be several possible positions for multiple bonding. Be sure to draw them all.
6. Add extra electrons in order to fill all valence shells.
7. Compute the formal charges for each atom in all possible structures.
8. (A special bit of advice.) Confirm the correct structure by guaranteeing that:
   a. each atom contains a filled valence shell.
   b. all valence electrons have been used.
   c. the lowest possible formal charges exist.
   d. formal charges of the same sign do not exist on adjacent atoms.

1. :Ö═Ö:

    1. 12 electrons are available.
    2. 16 electrons are needed.
    3. 4 electrons must be shared.
    4. 2 bonds
    5. O═O
    6. :Ö═Ö:
    7. :Ö═Ö:
    8. Check.

2. :Ö═Ö—Ö:
      ⊕  ⊖

    or

  :Ö—Ö═Ö:
   ⊖  ⊕

    1. 18 electrons are available.
    2. 24 electrons are needed.
    3. 6 electrons must be shared.
    4. 3 bonds
    5. O═O—O or O—O═O (equivalent)
    6. :Ö═Ö—Ö:
    7. :Ö═Ö—Ö:
    8. Check.

3.
```
      H
      |
  H—C—H
      |
      H
```

4. N̈═N═N̈
   ⊖ ⊕ ⊖

    1. 16 electrons are available, 5 from each N atom
       and 1 from the formation of the anion.
    2. 24 electrons are needed.
    3. 8 electrons must be shared.
    4. 4 bonds
    5. N═N═N or N≡N—N
    6. :N═N═N: or :N≡N—N̈:
    7. :N═N═N: or :N≡N—N̈:
      ⊖  ⊕  ⊖     ⊕ ②⊖
    8. The structure

        :N̈—N≡N:
        ②⊖  ⊕

    is improbable because it has a large formal charge
    on a single atom.

5. :F̈—Ö—F̈:

6.
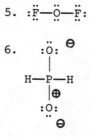
```
      :Ö:  ⊖
       |
   H—P—H
       |⊕
      :O:
       ⊖
```

7.  :F̈—N̈=N̈—F̈:    The structure

$$:\ddot{F}—\ddot{N}—\overset{\oplus}{\ddot{N}}=\overset{\ominus}{\overset{..}{\underset{}{F}}}:$$

is not probable, because the formal charges are not
necessary.

9.    :F̈:
      |
 :F̈—B—F̈:
      |⊖
      :F̈:

10.    H
       |⊕
  H—N—H
       |
       H

11.    :O:          The structure
       ‖
       N⊕
  :Ö⊖ ⊖Ö:

      or

is doubtful because there are only 6 electrons in the
valence shell of nitrogen and the formal charge of
nitrogen is larger than necessary.

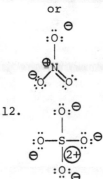

      or

      ..  ⊖
      :O:
       |
      ⊕N
  ⊖Ö⊖ ⊖Ö:

12.     :Ö:⊖
        |
 :Ö—S—Ö:⊖
 ⊖   |(2+)
     :O:⊖

13.  :N≡N:

14.  :Ö—C̈l:
     ⊖

15.  :N≡N—Ö:
     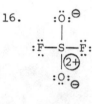
     ⊕  ⊖

The total number of valence electrons is 16. The structure

N̈=N=Ö̈
⊖  ⊕

is less probable because there are formal charges of opposite sign on adjacent atoms of the same element. The structure

:N̈—N≡O:
②⊖  ⊕  ⊕

is improbable because there are large formal charges on adjacent atoms.

16.      :Ö:⊖
          |
     :F̈—S—F̈:
         |②+
        :Ö: ⊖

17.   H   H
      |   |
     :N—N:
      |   |
      H   H

18.      :C̈l:
          |
     :C̈l—Sn—C̈l:
          |
         :C̈l:

19.  ⊖Ö̈
     :Ö—H

20.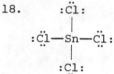

     or

     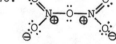

The total number of valence electrons is 40. The structures

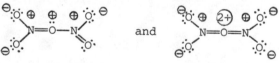

are improbable because in each there are positive formal charges on three adjacent atoms.

21.  ⊖:Ö—N̈=N—Ö:⊖

The total number of valence electrons is 24. The structure

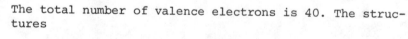

:Ö—N—N̈=Ö
   ⊖  ⊖

is improbable because the formal charges are localized at one end of the molecule.

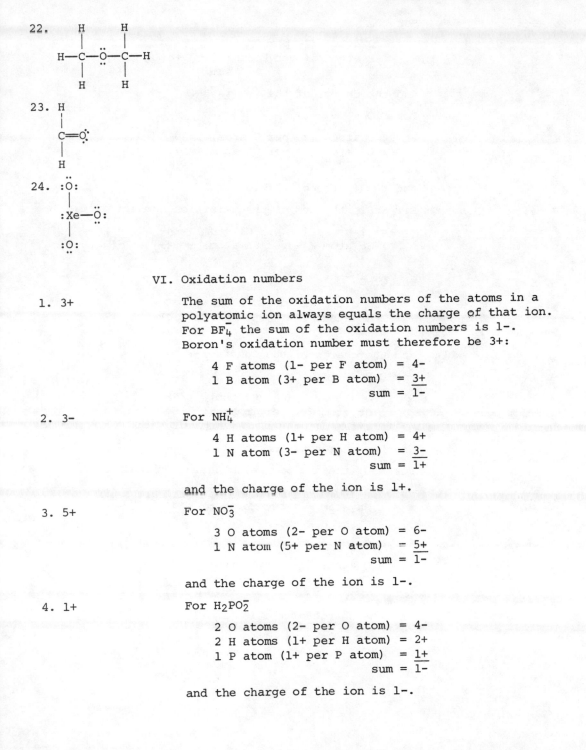

22.

```
     H       H
     |       |
 H — C — Ö — C — H
     |       |
     H       H
```

23.
```
 H
 |
 C = Ö
 |
 H
```

24.
```
    ··
   :O:
    |
 :Xe — Ö:
    |  ··
   :O:
    ··
```

VI. Oxidation numbers

1. 3+

The sum of the oxidation numbers of the atoms in a polyatomic ion always equals the charge of that ion. For $BF_4^-$ the sum of the oxidation numbers is 1−. Boron's oxidation number must therefore be 3+:

$$
\begin{array}{ll}
4 \text{ F atoms (1− per F atom)} & = 4- \\
1 \text{ B atom (3+ per B atom)} & = \underline{3+} \\
\text{sum} & = \overline{1-}
\end{array}
$$

2. 3−

For $NH_4^+$

$$
\begin{array}{ll}
4 \text{ H atoms (1+ per H atom)} & = 4+ \\
1 \text{ N atom (3− per N atom)} & = \underline{3-} \\
\text{sum} & = \overline{1+}
\end{array}
$$

and the charge of the ion is 1+.

3. 5+

For $NO_3^-$

$$
\begin{array}{ll}
3 \text{ O atoms (2− per O atom)} & = 6- \\
1 \text{ N atom (5+ per N atom)} & = \underline{5+} \\
\text{sum} & = \overline{1-}
\end{array}
$$

and the charge of the ion is 1−.

4. 1+

For $H_2PO_2^-$

$$
\begin{array}{ll}
2 \text{ O atoms (2− per O atom)} & = 4- \\
2 \text{ H atoms (1+ per H atom)} & = 2+ \\
1 \text{ P atom (1+ per P atom)} & = \underline{1+} \\
\text{sum} & = \overline{1-}
\end{array}
$$

and the charge of the ion is 1−.

5. 5+            For $PO_4^{3-}$

        4 O atoms (2- per O atom) = 8-
        1 P atom (5+ per P atom)  = 5+
                             sum = 3-

and the charge of the ion is 3-.

6. 7+            For $ClO_4^-$

        4 O atoms (2- per O atom)  = 8-
        1 Cl atom (7+ per Cl atom) = 7+
                         sum = 1-

and the charge of the ion is 1-.

7. 4+            For $SnCl_4$ the sum of the oxization numbers is 0.

        4 Cl atoms (1- per Cl atom) = 4-
        1 Sn atom (4+ per Sn atom)  = 4+
                       sum = 0

8. 6+            For $Cr_2O_7^{2-}$

        7 O atoms (2- per O atom)   = 14-
        2 Cr atoms (6+ per Cr atom) = 12+
                      sum = 2-

and the charge of the ion is 2-.

9. 6+            For $CrO_4^{2-}$

        4 O atoms (2- per O atom)  = 8-
        1 Cr atom (6+ per Cr atom) = 6+
                     sum = 2-

and the charge of the ion is 2-.

10. 4+           For $MnO_2$ the sum of the oxidation numbers is 0.

        2 O atoms (2- per O atom)  = 4-
        1 Mn atom (4+ per Mn atom) = 4+
                     sum = 0

11. 7+           For $MnO_4^-$

        4 O atoms (2- per O atom)  = 8-
        1 Mn atom (7+ per Mn atom) = 7+
                     sum = 1-

and the charge of the ion is -1.

12. 0            Any atom in a molecule of an element has an oxidation
                number of 0.

# SELF-TEST

Complete the test in 45 minutes.

I. Complete the following table:

| Formula of Compound | Name of Compound | Symbol of Element | Oxidation Number of Element |
|---|---|---|---|
| $SnCl_4$ | _____ | Sn | _____ |
| $As_2O_3$ | _____ | As | _____ |
| $Cu(NO_3)_2$ | _____ | N | _____ |
| _____ | manganese(IV) oxide | Mn | _____ |
| $Cr_2O_3$ | _____ | Cr | _____ |
| $NaC_2H_3O_2$ | _____ | Na | _____ |
| $N_2O_4$ | _____ | _____ | 4+ |
| $H_2SO_4$ | sulfuric acid | S | _____ |
| _____ | sodium chlorite | _____ | 3+ |

II. Answer each of the following:

_____
1. Which of the following elements has the greatest electron affinity?
   a. Na          c. I
   b. O           d. Cl

_____
2. Which of the following elements has the lowest first ionization energy?
   a. Na          c. I
   b. O           d. Cl

_____
3. Which of the following ions has the largest ionic radius?
   a. $Na^+$          c. $Ca^{2+}$
   b. $K^+$           d. $Ga^{3+}$

_____
4. Which of the following elements has the greatest electronegativity?
   a. Na          c. Sn
   b. Ca          d. F

_____
5. Which of the following elements has the smallest electronegativity?
   a. P           c. As
   b. S           d. Ge

IV. Which of the following is the best Lewis structure for NCCN? What is wrong with those structures that you did not choose?

a. :N≡C—C≡N:

b. :N̈—C≡C—N̈:

c. :N̈=C=C=N̈:

d. :N̈=C=C=N̈:

e. :N̈—C̈—C̈—N̈:

f. :N≡C=C̈—N̈:

# Molecular Geometry

OBJECTIVES

I. You should be able to demonstrate your knowledge of the following terms by defining them, describing them, and giving specific examples of them:

antibonding orbital [4.5]
aufbau order [4.5]
axial [4.3]
band [4.8]
bond order [4.5]
conductor [4.8]
degeneracy [4.5, 4.8]
diamagnetic [4.5]
dipole—dipole forces [4.7]
dipole moment [4.9]
dispersion forces [4.7]
electron pair repulsions [4.3]
equitorial [4.3]

homonuclear [4.5]
hybrid orbitals [4.4]
insulator [4.8]
ionic crystals [4.9]
irregular tetrahedron [4.3]
London forces [4.7]
metallic crystals [4.9]
molecular crystals [4.9]
molecular orbitals [4.5]
network crystals [4.9]
nonbonding pairs [4.3]
octahedron [4.3]
paramagnetic [4.5]
pi ($\pi$) orbital [4.5]
resonance [4.1]
resonance form [4.1]
sigma ($\sigma$) orbital [4.5]
square pyramid [4.3]
trigonal bipyramid [4.3]
trigonal pyramid [4.3]
VSEPR [4.3]

II. You should be able to draw resonance structures of molecules and ions.

III. You should be able to use VSEPR theory to predict molecular shapes.

IV. You should be able to determine whether or not a molecule is polar from its structure.

V. You should be able to determine the magnetic susceptibility and bond order of a diatomic molecule or ion.

VI. You should be able to determine boiling point and melting point trends for a series of related compounds.

VII. You should be able to rationalize observed molecular structures by drawing resonance forms.

VIII. You should be familiar with the properties and structures of crystalline solids.

EXERCISES      I. Draw resonance structures of a molecule of each of the following compounds. Include formal charges.

1. $COCl_2$: This compound, called phosgene or carbonyl chloride, is a poisonous gas. Carbon is the central atom in the compound.

2. $SO_3$: Sulfur trioxide is an atmospheric contaminant that is responsible for imparting the characteristic blue color to smoke. It is the acid anhydride of sulfuric acid, i.e., $SO_3$ reacts with water to form $H_2SO_4$:

$$SO_3(g) + H_2O(aq) \rightarrow H_2SO_4(aq)$$

3. OCCCO: Little chemistry is reported for this rare gas, which is called tricarbon dioxide.

4. ClCN: Chlorine cyanide or cyanogen chloride is the name of this poisonous, volatile liquid.

5. $NO_3^-$: This extremely common anion is the nitrate ion. Nitrogen is the central atom in the structure of this ion.

6. $O_3$: This compound, ozone, is a major reactant in photochemical smog production and is formed by the action of ultraviolet light on oxygen in the upper atmosphere.

7. NNO: The chemical name of this compound is dinitrogen monoxide or nitrous oxide. The compound is the anesthetic commonly referred to as "laughing gas." It is also used as a propellant gas in whipped cream cans.

II. Use the concept of electron pair repulsions, VSEPR, to predict the geometric configuration of a molecule of each of the following compounds:

1. $XeF_4$, xenon tetrafluoride

2. $AsF_5$, arsenic pentafluoride

3. $SnF_4$, tin tetrafluoride or tin(IV) fluoride

4. $IF_2^+$, the iodine difluoride cation

5. $ClF_3$, chlorine trifluoride

6. $H_3O^+$, the hydronium ion

7. $I_3^-$, the triiodide ion

8. $SnBr_6^{2-}$, the tin(IV) hexabromide anion

9. $SF_4$, sulfur tetrafluoride

III. Draw the most probable Lewis structure for each of the following compounds. Use VSEPR theory to predict the molecular shape.

1. $PH_3$
2. $SnCl_4$
3. $SnCl_2$
4. $O_3$
5. $N_3^-$

6. $CO_2$
7. $SO_2$
8. HCN
9. FNO
10. $H_2CO$

IV. Predict the orbital hybridization of the central atom in each of the molecules described in Sections II and III.

V. Write the electronic configuration of each of the following homopolar, diatomic molecules and ions. Determine the bond order of each and determine whether the molecule or ion is paramagnetic or diamagnetic. Also draw the Lewis structure of each.

1. NO          2. $CN^-$          3. $O_2^+$

VI. Predict the geometric configuration of a molecule of each of the following compounds. Describe the bonding between all atoms in the molecule. Remember that the molecular shapes will only be approximations based on the theories with which you are now familiar.

1. $H_2CO$, formaldehyde
2. $C_6H_6$, a six-carbon ring compound, benzene
3. $CCl_4$, carbon tetrachloride
4. HCCH, acetylene
5. $SO_2$, sulfur dioxide
6. FNNF, dinitrogen difluoride
7. HF, hydrogen fluoride

VII. From the following list of molecules and atoms, described by their Lewis structures, choose the ones that answer the question. A molecule or atom may be used to answer more than one question.

a. H—Ö—H

b. H—N̈—H
       |
       H

c. H—C̈l:

d.        H  H
          |  |
      H—C—C—H
          |  |
          H  H

e.

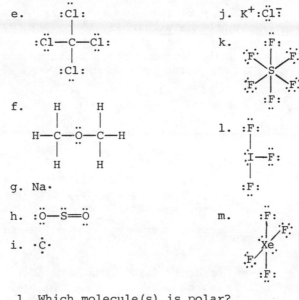

```
        :Cl:
         ..
  :Cl—C—Cl:
   ..  |  ..
        :Cl:
         ..
```

f.

```
    H       H
    |       |
    ..      ..
H—C—O—C—H
    |       |
    H       H
```

g. Na·

h. :Ö—S̈=Ö

i. ·Ċ·

j. K⁺:C̈l:⁻

k.

l.

m.

1. Which molecule(s) is polar?
2. Which molecule(s) is nonpolar?
3. Which molecule(s) contains one or mote atoms that probably use $sp^3$ hybrid orbitals in bonding?
4. Which molecule(s) contain π bonds?
5. Which atom(s) forms metallic bonds?
6. Which molecule(s) forms ionic crystals?
7. Which molecule(s) contain seven σ bonds?
8. Which molecule(s) or atom(s) contains unpaired electrons?
9. Which molecule(s) is planar?
10. Which molecule(s) does not contain nonbonding pairs of electrons?

**ANSWERS TO EXERCISES**

I. Resonance structures
   See Section 4.1 of your text.

   1.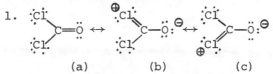

   (a)        (b)        (c)

   Three resonance structures (a, b, and c) can be drawn. The structure of a $COCl_2$ molecule can be described as a resonance hybrid, or a weighted average, of the three structures. Structure (a) is more important than either (b) or (c), i.e., the actual structure of a $COCl_2$ molecule contains

more of the expected character structure (a) than
of that of either (b) or (c). If you had been asked
to draw a single Lewis structure to describe the molecule, structure (a) would be correct, because it
most closely approximates the structure of the molecule.

2.

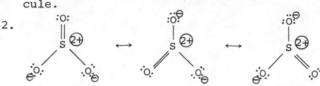

In sulfur trioxide all S—O bonds are of equal
length. The actual structure is an average of the
three resonance forms, each contributing equally.

3. $\ddot{O}=C=C=C=\ddot{O} \leftrightarrow :\overset{\oplus}{\ddot{O}}\equiv C-C\equiv C-\overset{\ominus}{\ddot{O}}: \leftrightarrow :\overset{\ominus}{\ddot{O}}-C\equiv C-C\equiv\overset{\oplus}{O}:$

        (a)               (b)           (c)

Structure (a) is the main contributor to the actual
structure of tricarbon dioxide.

4. $:\ddot{C}l—C\equiv N: \leftrightarrow \overset{..}{C}l\overset{\oplus}{=}C=\overset{..}{\overset{\ominus}{N}}$

      (a)           (b)

Structure (a) is a more important contributor to
the actual structure than structure (b).

5.

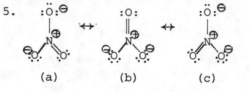

    (a)      (b)      (c)

All three structures contribute equally to the
actual structure of the nitrate ion.

6.

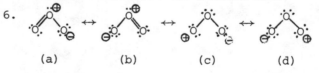

    (a)      (b)      (c)      (d)

Four structures can be drawn for the ozone molecule.
Structures (a) and (b) are expected to contribute
most to the actual structure. Structures (c) and (d)
do not obey the octet rule.

7. $\overset{\ominus}{N}=\overset{..}{N}\overset{\oplus}{=}\ddot{O} \leftrightarrow :N\equiv\overset{\oplus}{N}—\overset{..}{\overset{\ominus}{O}}:$

      (a)          (b)

Both are important contributors to the actual
structure of nitrous oxide. A third structure

②$:N—N≡O:$

does not contribute appreciably.

II. Electron pair repulsions and molecular geometry

See Section 4.3 of your text.

1. square planar
   configuration

To predict the geometry of a molecule from a consideration of electron pair repulsions, we can use the following method if the molecule contains no double bonds:

Electron-pair repulsion method:
(a) Determine the number of electrons in the valence
    shell of the neutral central atom. If the molecule
    is an ion, account for its charge by adding or
    subtracting electrons in the valence shell of the
    central atom.
(b) Determine the total number of electrons contrib-
    uted to the central atom's valence shell by the
    atoms directly bonded to the central atom.
(c) Divide the sum of (a) and (b) by 2 to determine
    the number of electron pairs.
(d) Distribute the electron pairs in the valence shell
    of the central atom such that they are as far away
    from each other as possible. Such orientations of
    electron pairs are given in Table 4.1 of the study
    guide.

TABLE 4.1  Orientations of Electron Pairs about the Center
           of the Central Atom of a Molecule

| Number of Electron Pairs | Orientation | |
|---|---|---|
| 2 | linear | $180°$ |
| 3 | triangular planar | $120°$ |
| 4 | tetrahedral | $109°$ |
| 5 | trigonal bipyramidal | $90°$  $120°$ |
| 6 | octahedral | $90°$  $90°$ |

(e) Bond the peripheral atoms to the central atom
such that minimum repulsion occurs. Molecular
shapes predicted from a consideration of minimum electron pair repulsions are given in Table
4.2 of the study guide.

Using the electron-pair repulsion method to predict the
geometry of $XeF_4$, we find:

(a) The total number of valence electrons in a neutral atom of xenon is 8. Since $XeF_4$ is uncharged,
no electrons must be subtracted from or added to
the valence shell of xenon.

(b) One electron is donated to the valence shell of
xenon by each fluorine atom. The total number of
electrons donated is therefore 4.

(c) 8 electrons + 4 electrons = 12 electrons
12 electrons/2 = 6 electron pairs

(d) According to the information in Table 4.1 of the
study guide, we predict the electron pairs to
have an octahedral orientation about the center
of the xenon atom.

(e) According to the information in Table 4.2 of the
study guide, we predict the molecule to have a
square planar shape.

2. trigonal
   bipyramidal

Electron-pair repulsion method
(a) Five valence electrons are in the valence shell
of arsenic.
(b) Five electrons are contributed by the peripheral
atoms, one by each fluorine atom.
(c) There is a total of five electron pairs.

$(5 + 5)/2 = 5$

(d) Electron pairs are distributed such that they
have a trigonal bipyramidal orientation.
(e) We predict the shape of the molecule is that of
a trigonal bipyramid.

3. tetrahedral

Electron-pair repulsion method
(a) Four electrons are in the valence shell of tin.
(b) Four electrons are contributed by the peripheral
(c) There is a total of four electron pairs.
(d) Electron pairs are distributed such that they
have a tetrahedral orientation.
(e) We predict the shape of the molecule to be that
of a tetrahedron.

TABLE 4.2  Number of Electron Pairs in the Valence Shell of the Central
Atom and Molecular Shape

| Total | Bonding | Nonbonding | Shape of Molecule or Ion | Examples |
|-------|---------|------------|--------------------------|----------|
| | Number of Electron Pairs | | | |
| 2 | 2 | 0 | linear | $HgCl_2$, $CuCl_2^-$ |
| 3 | 3 | 0 | triangular planar | $BF_3$, $HgCl_3^-$ |
| 3 | 2 | 1 | angular | $SnCl_2$, $NO_2^-$ |
| 4 | 4 | 0 | tetrahedral | $CH_4$, $BF_4^-$ |
| 4 | 3 | 1 | trigonal pyramidal | $NH_3$, $PF_3$ |
| 4 | 2 | 2 | angular | $H_2O$, $ICl_2^+$ |
| 5 | 5 | 0 | trigonal bipyramidal | $PCl_5$, $SnCl_5^-$ |
| 5 | 4 | 1 | distorted tetrahedral | $TeCl_4$, $IF_4^+$ |
| 5 | 3 | 2 | T shaped | $ClF_3$ |
| 5 | 2 | 3 | linear | $XeF_2$, $ICl_2^-$ |
| 6 | 6 | 0 | octahedral | $SF_6$, $PF_6^-$ |
| 6 | 5 | 1 | square pyramidal | $IF_5$, $SbF_5^{2-}$ |
| 6 | 4 | 2 | square planar | $BrF_4^-$, $XeF_4$ |

4. angular

(a) There are seven electrons in the valence shell of a neutral iodine atom. Since the molecule has 1+ charge, we must subtract one electron from the valence shell of the iodine atom. Thus, the valence shell of the iodine atom has six electrons.

(b) Two electrons are contributed by the peripheral atoms, one by each fluorine atom.

(c) There is a total of four electron pairs.

(d) Electron pairs are distributed such that they have a tetrahedral orientation.

(e) We predict the molecule to have an angular shape.

5. T shaped

6. trigonal pyramidal

7. linear

8. octahedral

9. tetrahedral

III. VSEPR theory continued:
To predict molecular shapes by drawing Lewis structures use the following procedure:
(a) Draw the structure using the method described in Chapter 3.
(b) Count the number of lone pairs on the central atom.
(c) Count the number of peripheral atoms bonded to the central atom.
(d) Distribute the sum of (b) and (c) around the central atom so that they are as far away from each other as possible. See Table 4.1 in the study guide.
(e) Add the double bonds.

1. trigonal pyramidal

Applying these rules to $PH_3$:
(a) H—P̈—H
          |
          H
(b) One lone pair
(c) Three peripheral atoms
(d) Tetrahedral distribution

(e)

2. tetrahedral

3. angular

Experimental evidence shows that the double bond does not exist. The correct angular shape is predicted by this approach anyway. The actual structure of $SnCl_2$ is discussed in Section 4.3 of your text. Although the actual structure is

it would be a fair assumption to predict the possibility of some double bond character in the absence of evidence to the contrary. Your instructor will guide you by choosing molecules in which the experimental evidence and theory are consistent.

4. angular

Applying the rules outlined in problem 1 in this section of the study guide:
   (a) $\ddot{O}=\ddot{O}-\ddot{O}:$
   (b) One lone pair
   (c) Two peripheral atoms
   (d) Triangular planar distribution

   (e)

5. linear

N=N=N $^\ominus$

   (a) $:N=N=N:^\ominus$
   (b) No lone pairs
   (c) Two peripheral atoms
   (d) Linear distribution
   (e) N=N=N$^\ominus$

6. linear

O=C=O

7. angular

8. linear

H—C≡N

9. angular

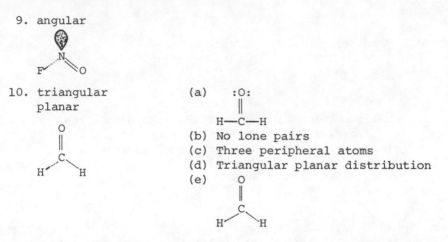

10. triangular
    planar

(a)   :O:
         ‖
      H—C—H
(b)  No lone pairs
(c)  Three peripheral atoms
(d)  Triangular planar distribution
(e)      O
         ‖
         C
      H⁄   ⁀H

IV. Hybrid orbitals and molecular geometry

1. (II)
   $d^2sp^3$

2. (II)
   $dsp^3$

3. (II)
   $sp^3$

4. (II)
   $sp^3$

5. (II)
   $dsp^3$

6. (II)
   $sp^3$

7. (II)
   $dsp^3$

8. (II)
   $d^2sp^3$

9. (II)
   $sp^3$

1. (III)
   $sp^3$

2. (III)
   $sp^3$

3. (III)
   $sp^2$

4. (III)
   $sp^2$

Normally, experimental evidence is needed to predict the types of hybrid orbitals involved in bonding. In the absence of such evidence the hybrid orbitals can be determined from the assumed molecular shape and the number of nonbonded pairs of electrons. The total number of nonbonding electron pairs plus the number of peripheral atoms (atoms bonded to the central atom) always equals the total number of orbitals forming the hybrid. For instance, a $d^2sp^3$ orbital is a hybrid of two $d$ orbitals plus one $s$ orbital plus three $p$ orbitals, for a total of six orbitals. Thus any combination of nonbonding pairs plus peripheral atoms which totals six might well indicate that the central atom is $d^2sp^3$ hybridized. Refer to Table 4.3 in the study guide to answer all questions in this section.

TABLE 4.3   Geometry and Hybridization of Atomic Orbitals

| Number of Nonbonding Electrons Pairs + Number of Atoms Bonded to the Central Atom | Type of Hybrid Orbitals Used by the Central Atom | Geometry of Hybrid Orbitals about the Center of the Central Atom |
|---|---|---|
| 2 | $sp$ | linear |
| 3 | $sp^2$ | triangular planar |
| 4 | $sp^3$ | tetrahedral |
| 5 | $dsp^3$ | trigonal pyramidal |
| 6 | $d^2sp^3$ | octahedral |

5. (III)
   $sp$

6. (III)
   $sp$

7. (III)
   $sp^2$

8. (III)
   $sp$

9. (III)
   $sp^2$

10. (III)
    $sp^2$

## V. Molecular orbitals

It is imperative that you read your class notes very meticulously to ascertain the depth to which your instructor wishes you to master this material.

1. In an NO molecule 15 electrons are distributed as described in your text:

$$(\sigma_{1s})^2(\sigma_{1s}^*)^2(\sigma_{2s})^2(\sigma_{2s}^*)^2(\sigma_{2p})^2(\pi_{2p})^4(\pi_{2p}^*)^1$$

The bond order is $2\frac{1}{2}$: bond order

$$= \frac{10 \text{ bonding electrons } - \text{ 5 antibonding electrons}}{2}$$

$$= 2\frac{1}{2}$$

A molecular orbital energy-level diagram for the higher-energy molecular orbitals of NO can be drawn:

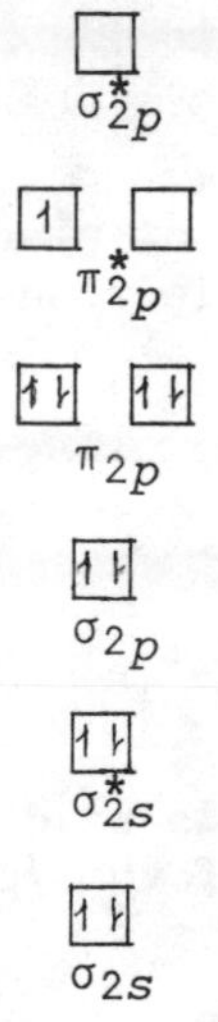

$$\boxed{\phantom{\uparrow\downarrow}}$$
$\sigma_{2p}^*$

$$\boxed{\uparrow}\quad\boxed{\phantom{\uparrow}}$$
$\pi_{2p}^*$

$$\boxed{\uparrow\downarrow}\quad\boxed{\uparrow\downarrow}$$
$\pi_{2p}$

$$\boxed{\uparrow\downarrow}$$
$\sigma_{2p}$

$$\boxed{\uparrow\downarrow}$$
$\sigma_{2s}^*$

$$\boxed{\uparrow\downarrow}$$
$\sigma_{2s}$

From this diagram we radily see that there is one unpaired electron in an NO molecule. Thus, an NO molecule is paramagnetic. Lewis structures do not give an adequate description of the NO molecule:

$$:\overset{..}{N}=\overset{..}{O}: \leftrightarrow :\overset{..}{N}=\overset{..}{O}: \leftrightarrow :\overset{..}{N}=\overset{..}{O}:$$

2. In a $CN^-$ ion there are 14 electrons to be distributed in molecular orbitals:

$$(\sigma_{1s})^2 (\sigma_{1s}^*)^2 (\sigma_{2s})^2 (\sigma_{2s}^*)^2 (\pi_{2p})^4 (\pi_{2p})^2$$

The bond order is 3. A molecular orbital energy-level diagram shows that there are no unpaired electrons in a $CN^-$ ion. A $CN^-$ ion is therefore diamagnetic. The Lewis structure

$$:C\equiv N: \overset{\ominus}{}$$

gives an adequate description of the bond ordrr of the molecule.

3. In an $O_2^+$ ion there are 15 electrons to be distributed in molecular orbitals:

$$(\sigma_{1s})^2 (\sigma_{1s}^*)^2 (\sigma_{2s})^2 (\sigma_{2s}^*)^2 (\sigma_{2p})^2 (\pi_{2p})^4 (\pi_{2p}^*)^1$$

A molecular orbital diagram shows that there is one unpaired election in the $O_2^+$ ion. An $O_2^+$ ion therefore should be paramagnetic. Lewis structures do not give an adequate description of the bond order of the ion:

$$:\overset{..}{O}=\overset{..}{O}: \leftrightarrow :\overset{..}{O}=\overset{..}{O}: \leftrightarrow :\overset{..}{O}\equiv O:$$

VI. Geometric configurations of molecules

1. triangular planar

The Lewis structure is

$$
\begin{array}{c}
:O: \\
\parallel \\
H-C-H
\end{array}
$$

According to the concept of electrion pair repulsion, the σ bonding pairs of electrons are distributed about the center of the carbon atom in the configuration of a plane triangle:

The central carbon atom is said to be $sp^2$ hybridized. The $2s$ orbital and two of the $2p$ orbitals of carbon form the $sp^2$ hybrid orbitals of carbon while the

remaining *p* orbital of carbon overlaps a *p* orbital of
oxygen to form a π bond:

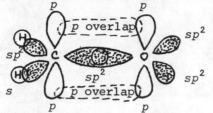

2. hexagonal
   planar

There are two resonance forms of benzene:

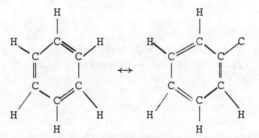

According to the concept of electron pair repulsion,
the σ bonding pairs of electrons are distributed about
the centers of the appropriate carbon atoms in the configuration of a plane triangle:

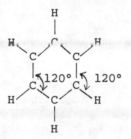

Each carbon is said to be $sp^2$ hybridized. The $2s$ orbital
and two of the $2p$ orbitals of carbon form the $sp^2$ hybrid
orbitals of carbon while the remaining *p* orbital of carbon enters into   bonding (see Figure 4.1).

3. tetrahedral

The Lewis structure is

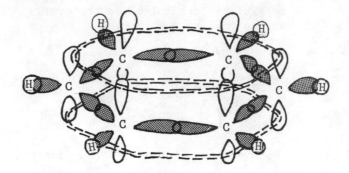

FIGURE 4.1  The $sp^2$ hybrid orbitals of the carbon atoms are shaded, and the unhybridized $p$ orbitals are not. The $p$ orbital overlap shown on a decreased scale is indicated by a dashed line.

According to the concept of electron pair repulsion, a tetrahedral geometry is predicted:

$$
\begin{array}{c}
Cl \\
| \\
Cl \text{---} C \text{---} Cl \\
| \\
Cl
\end{array}
$$

The carbon is said to be $sp^3$ hybridized.

4. linear

The Lewis structure is

H—C≡C—H

Each carbon has two atoms bonded to it and no nonbonding pairs of electrons. According to the concept of electron pair repulsion, the molecule has a linear shape. Each carbon is said to be $sp$ hybridized and forms two σ bonds.

a σ bond formed by the overlap of an $sp$ hybrid orbital of each carbon

a σ bond formed by the overlap of an orbital of hydrogen and an $sp$ hybrid orbital of carbon

Each carbon also forms two π bonds. These bonds are formed by the overlap of the unhybridized $p$ orbitals of each carbon.

one σ bond and two π bonds

5. angular

The resonance forms of $SO_2$ are

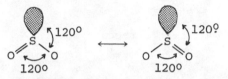

Sulfur is bonded to two atoms and has one nonbonding pair of electrons. According to electron pair repulsion, the σ and nonbonding electron pairs are distributed about the center of the sulfur atom in the configuration of a plane triangle:

It is difficult to predict the bonding of the oxygen atoms, because the π bond is not localized between any two atoms:

6. planar
   bent

The Lewis structure is

:F̈—N̈=N̈—F̈:

According to the concept of electron pair repulsion, the σ and nonbonding pairs of electrons are distributed about each nitrogen atom in a planar triangular configuration:

N=N structure with F atoms

Each nitrogen is said to be $sp^2$ hybridized. A σ bond between the nitrogen atoms is formed from the overlap of these $sp^2$ hybrid orbitals. Each nitrogen also has one $sp^2$ orbital overlapping with a fluorine orbital to form an N—F σ bond. The remaining $sp^2$ nitrogen orbital contains a nonbonding pair of electrons. The unhybridized $p$ orbitals overlap to form a π bond:

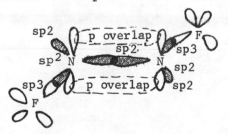

It is difficult to prove the type of hybrid orbitals, if any, associated with the fluorine atom, because the structure of the molecule would be little affected by such hybridization. The F atoms are drawn with $sp^3$ hybrid orbitals in this problem. A structure in which the fluorine atoms are on the same side of the molecule is also possible:

7. linear

The Lewis structure is

H : F̈:

The fluorine is said to be $sp^3$ hybridized. A tetrahedral arrangement of the bonding pair of electrons and the nonbonding pairs of electrons provides the most effective separation of electron pairs.

VII. Structure of molecules

1. a, b, c, f, h, j, l

The water molecule is angular. The individual dipoles combine to make the molecule a dipole. The direction of an individual dipole is indicated by an arrow, the arrow pointing toward the negative end of the dipole.

Ammonia is triangular pyramidal. The individual dipoles combine to make the molecule a dipole.

The individual dipoles cancel in a molecule that is symmetrical about a central point:

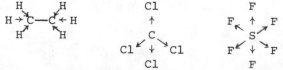

(nonbonding electron pairs of Cl omitted for clarity)

(nonbonding electron pairs of F omitted for clarity)

A $(CH_3)_2O$ molecule, like water, is angular. The individual dipoles combine to make the molecule a dipole.

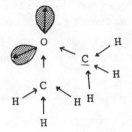

Because the molecule $SO_2$ is angular, the individual dipoles do not cancel:

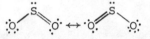

The $IF_3$ molecule is T shaped and the individual dipoles do not cancel:

     (nonbonding electron pairs of fluorines omitted for clarity)

The individual bond dipoles of $XeF_4$ totally cancel each other and the molecule is nonpolar.

     (nonbonding electron pairs of fluorines omitted for clarity)

2. d, e, k, m          See Section 4.6 of your text.

3. a, b, c, d,         See Section 4.5 of your text.
   e, f
    (a) The oxygen atom probably uses $sp^3$ hybrid orbitals.
    (b) The nitrogen atom probably uses $sp^3$ hybrid orbitals.
    (c) The chlorine atom *possibly* uses $sp^3$ hybrid orbitals.
    (d) Each carbon atom probably uses $sp^3$ hybrid orbitals.
    (e) The carbon atom probably uses $sp^3$ hybrid orbitals, and the chlorine atom *possibly* uses $sp^3$ hybrid orbitals.
    (f) The oxygen atom and the carbon atoms each probably use $sp^3$ hybrid orbitals.

4. h                   See Sections 4.4 and 4.5 of your text.

5. g                    See Section 4.9 of your text.

6. j                    See Section 4.9 of your text.

7. d                    See Sections 4.4 and 4.5 of your text.

8. g, i                 See Section 4.4 of your text.
                        g: An atom of sodium has a single electron in the 2s
                           subshell.
                        i: An atom of carbon has two unpaired electrons in the
                           2p subshell.

9. a, c, h, j.          See Section 4.3 of your text.
   l, m

10. d


SELF-TEST               Complete the test in 30 minutes.

   1. Complete the table:

| Formula | Number of Electron Pairs | | Shape of Molecule or Ion |
| | Bonding | Nonbonding | |
| --- | --- | --- | --- |
| $SCl_2$ | _____ | _____ | _____ |
| $XeF_4$ | _____ | _____ | _____ |
| $AlH_4^-$ | _____ | _____ | _____ |
| $TeCl_4$ | _____ | _____ | _____ |
| $SeF_5^-$ | _____ | _____ | _____ |

   2. Crystalline $PCl_5$ consists of an ionic lattice of
      $PCl_4^+$ and $PCl_6^-$ ions. In the vapor and liquid states,
      however, the compound exists as $PCl$  molecules. What
      type of hybrid orbitals does P employ in each of
      these species, and what is the geometry of the mole-
      cule or ion?

|  | Hybrid Orbitals | Geometric Shape |
| --- | --- | --- |
| $PCl_5$ | _____ | _____ |
| $PCl_4^+$ | _____ | _____ |
| $PCl_6^-$ | _____ | _____ |

   3. Complete the following table. The σ orbitals include
      only the σ2s and σ2p orbitals, and the σ* orbitals
      include only the σ*2s and σ*2p orbitals.

| Molecule | Total Number of Electrons in | | | | Bond Order | Number of unpaired Electrons |
|---|---|---|---|---|---|---|
|  | σ Orbitals | σ* Orbitals | π Orbitals | π* Orbitals |  |  |
| Be$_2$ | _____ | _____ | _____ | _____ | _____ | _____ |
| B$_2$ | _____ | _____ | _____ | _____ | _____ | _____ |
| N$_2$ | _____ | _____ | _____ | _____ | _____ | _____ |
| O$_2$ | _____ | _____ | _____ | _____ | _____ | _____ |
| NO$^+$ | _____ | _____ | _____ | _____ | _____ | _____ |

4. The thiocyanate ion, SCN$^-$, is linear and the atoms are arranged in the order given. Draw Lewis structures for the three resonance forms of the ion *complete with formal charges*.

# Chemical Equations and Quantitative Relations

OBJECTIVES

I. You should be able to demonstrate your knowledge of the following terms by defining them, describing them, or giving specific examples of them:

atomic mass unit [5.1]
Avogadro's number [5.1]
Born-Haber cycle [5.10]
calorimeter [5.6]
combustion [5.8]
empirical formula [5.2]
endothermic [5.7]
enthalpy [5.7
enthalpy of combustion [5.7]
enthalpy of formation [5.9]
enthalpy of reaction [5.7]
exothermic [5.7]
heat capacity [5.6]
joule [5.6]

72

lattice energy [5.10]
law of conservation of mass [5.4]
law of constant composition [5.3]
law of Hess [5.8]
limiting reagent [5.5]
mole [5.1]
molecular formula [5.2]
percent yield [5.5]
specific heat [5.6]
standard state [5.9]
thermochemistry [5.6]
true formula [5.2]

II. You should be able to perform calculations to obtain numbers of moles and molecules, percent composition, empirical formulas, and molecular formulas.

III. You should be able to balance chemical equations and use them in stoichiometric calculations.

IV. You should be able to compute percent yields from actual yields and theoretical yields. You should be able to calculate theoretical yields.

V. You should be able to calculate enthalpies of reaction and heat capacities.

VI. You should be able to use the law of Hess and to construct Born-Haber cycles for thermochemical calculations.

## UNITS, SYMBOLS, MATHEMATICS

I. The factor-label method, which is also referred to as dimensional analysis and sometimes a logic chain, should be used in solving scientific problems. Frequently you may already use such a method, but you may not have formalized your thoughts in precisely this way. For instance, if you were to determine the number of eggs in 2 dozen, you would immediately answer 24. Including the labels we would write:

$$2 \text{ dozen} \left( \frac{12 \text{ eggs}}{1 \text{ dozen}} \right) = 24 \text{ eggs}$$

Note that the label "dozen" cancels from the equation and the label "eggs" remains. Labels are cancelled from the numerator and the denominator in exactly the same way that numbers are cancelled. In a similar way you might be asked to determine the total weight of tires discarded each year if the average tire weighs 30 pounds and 200 million are discarded:

$$? \text{ lb} = 200 \times 10^6 \text{ tires} \left(\frac{30 \text{ lb}}{1 \text{ tire}}\right)$$
$$= 6000 \times 10^6 \text{ lb}$$
$$= 6.0 \times 10^9 \text{ lb}$$

The question is "How many pounds?" so write "? lb =," and continue the problem to obtain the proper label. *Each factor must be an equivalency.* Here 1 tire equals 30 pounds. Using the original information and the fact that 2.20 lb = 1.00 kg, compute the number of kilograms of tires discarded each year:

$$? \text{ kg} = 200 \times 10^6 \text{ tires} \left(\frac{30 \text{ lb}}{1 \text{ tire}}\right)\left(\frac{1.00 \text{ kg}}{2.20 \text{ lb}}\right)$$
$$= 2.73 \times 10^9 \text{ kg}$$

If 25 percent of the rubber is recoverable for use as blacktop material, how many grams of tires per year could be recycled?

$$? \text{ g recycled}$$
$$= 200 \times 10^6 \text{ tires} \left(\frac{30 \text{ lb}}{1 \text{ tire}}\right)\left(\frac{1000 \text{ g}}{2.20 \text{ lb}}\right)\left(\frac{25 \text{ g recycled}}{100 \text{ g}}\right)$$
$$= 6.8 \times 10^{11} \text{ g recycled}$$

Observe that the procedure is the same as the stepwise analysis, except that intermediate answers are not recorded. Note the progressive changes in units:

$$? \text{ g recycled} = 200 \times 10^6 \text{ tires} \left(\frac{30 \text{ lb}}{\text{tire}}\right)\left(\frac{1 \text{ kg}}{2.20 \text{ lb}}\right)\left(\frac{1000 \text{ g}}{1 \text{ kg}}\right)\left(\frac{25 \text{ recycled}}{100}\right)$$
$$\phantom{xxxxxxxxxxxxxx} \text{lb} \rightarrow \quad \text{kg} \rightarrow \quad \text{g} \rightarrow \quad \text{g recycled}$$

## EXERCISES

I. Work the following problems. Recall that a factor is meaningless without corresponding units, the label.

1. Most iron is obtained from $Fe_2O_3$. What is the percentage of iron (by weight) in $Fe_2O_3$?

2. Gibbsite, $Al_2O_3 \cdot 3H_2O$, is a naturally occurring material from which aluminum is produced.*

---

*Many minerals and other materials are combinations of two or more compounds in specific ratios. Gibbsite is a combination of aluminum oxide, $Al_2O_3$, and water in a 1:3 ratio.

3. Chemical analyses are often performed to determine the percent composition of a pure material, and from the information the empirical formula can be determined. An organic material is analyzed and found to contain 75.92% C, 17.71% N, and 6.37% H. Determine the empirical formula of the compound.

4. Tritopine, an alkaloid isolated from opium, has been found to be 74.0% C, 7.90% H, 14.0% O, and 4.10% N by weight. What is the empirical formula? The molecular weight is known to be 682 g/mol. What is the molecular formula?

5. Commercially iron is obtained from the reduction of hematite, $Fe_2O_3$:

$$Fe_2O_3 + CO \rightarrow Fe + CO_2$$

How many grams of iron can be obtained from 5.24 g of $Fe_2O_3$?

6. If only 2.04 g of Fe is obtained from the sample described in problem 5 of this section, what is the percent yield of the reduction process?

7. How many grams of iron can be obtained by the reaction of 2.78 g of CO with excess $Fe_2O_3$? (See problem 5 of this section.)

8. The carbon monoxide needed for producing iron by the reaction given in problem 5 of this section is obtained by the reaction of oxygen with carbon, which is in the form of coke:

$$C + O_2 \rightarrow CO$$

How much iron can be obtained from 1.47 g of C?

9. Refer to problem 5 of this section and answer the following:
   (a) How much iron can be obtained from 4.02 g of $Fe_2O_3$ and 1.78 g of CO?
   (b) What are the masses of all compounds after the reaction is complete?

10. Ethyl alcohol, $C_2H_5OH$, is a product of the fermentation of sugars. In the presence of the enzyme invertase, sucrose undergoes conversion into invert sugar.*

---

*A catalyst such as an enzyme is not written in the balanced equation of a reaction; rather, it is written over the reaction arrow of the equation. Zymase increases the rate of conversion of sucrose to alcohol such that the conversion proceeds at a reasonable rate.

$$C_{12}H_{22}O_{11} + H_2O \xrightarrow{\text{invertase}} 2C_6H_{12}O_6$$

The invert sugar is converted into ethyl alcohol and carbon dioxide by the enzyme zymase. Usually only half the sucrose is converted into alcohol, a 50 percent yield.

$$C_6H_{12}O_6 \xrightarrow{\text{zymase}} 2C_2H_5OH + 2CO$$

How much ethyl alcohol, $C_2H_5OH$, can be obtained from 171 g of sucrose, $C_{12}H_{22}O_{11}$?

II. Use the data in the following table when needed to solve the thermochemical problems in this section. In the space provided write the balanced equation that corresponds to the listed enthalpy of formation.

### Enthalpies of Formation at 25°C

| Molecule | $\Delta H_f$ (kJ) | Equation |
|---|---|---|
| CO (g) | -110.5 | |
| $CO_2$ (g) | -393.7 | |
| $Fe_2O_3$ (s) | -822.2 | |
| HCl (g) | -92.5 | |
| $H_2O$ (g) | -241.8 | |
| $H_2O$ (l) | -285.8 | |
| $H_2SO_4$ (l) | -811.3 | |
| $I_2$ (g) | +62.3 | |
| $NH_3$ (l) | -66.9 | |
| NO (g) | +90.4 | |
| $NO_2$ (g) | +52.7 | |
| $O_3$ (g) | +142.3 | |
| $PCl_3$ (g) | -306.3 | |
| $PCl_5$ (g) | -399.2 | |
| $POCl_3$ (g) | -592.0 | |
| $SO_2$ (g) | -297.0 | |
| $SO_3$ (g) | -395.0 | |

1. What is the enthalpy change at 25°C for the following reaction?

   $SO_3(g) + H_2O(l) \rightarrow H_2SO_4(l)$

2. What is the enthalpy change at 25°C for the following reaction?

   $PCl_5(g) + H_2O(g) \rightarrow POCl_3(g) + 2HCl(g)$

3. The enthalpy of combustion of pentane, $C_5H_{12}(g)$, to $CO_2(g)$ and $H_2O(l)$ is -3536.3 kJ/mol. What is the enthalpy of formation of pentane at 25°C?

4. A 12.45 g sample of $P_4O_{10}(s)$ is reacted with a stoichiometrically equivalent quantity of water in a vessel placed in a calorimeter containing 950.0 g of water. The temperature of the calorimeter and its contents increases from 22.815°C to 26.885°C. If the calorimeter has a heat-absorbing capacity equal to that of 165.0 g $H_2O$, what is the enthalpy change for the following reaction?

   $P_4O_{10}(s) + 6H_2O(l) \rightarrow 4H_3PO_4(aq)$

   Assume the specific heat of water to be 4.184 J/g°C.

5. Calculate the lattice energy of potassium chloride from the following data. For potassium the ionization energy is 414 kJ/mol, and the enthalpy of sublimation if 88 kJ/mol. For chlorine the dissociation energy is 243 kJ/mol, and the electron affinity is 368 kJ/mol. The enthalpy of formation of KCl is 432 kJ/mol.

## ANSWERS TO EXERCISES

### I. Chemical calculations

1. 69.9434% Fe
   [5.3]

The formula indicates the ratio of moles. From this the ratio of masses must be calculated to determine percent composition. In 1 mole of $Fe_2O_3$ there would be how many grams of Fe and how many grams of O?

$$? \text{ g Fe/1 mol } Fe_2O_3 = \left(\frac{55.847 \text{ g Fe}}{1 \text{ mol Fe}}\right)\left(\frac{2 \text{ mol Fe}}{1 \text{ mol } Fe_2O_3}\right)$$

$$= 111.694 \text{ g Fe/1 mol } Fe_2O_3$$

$$? \text{ g O/1 mol } Fe_2O_3 = \left(\frac{15.9994 \text{ g O}}{1 \text{ mol O}}\right)\left(\frac{3 \text{ mol O}}{1 \text{ mol } Fe_2O_3}\right)$$

$$= 47.9982 \text{ g O/1 mol } Fe_2O_3$$

The mass of 1 mole of $Fe_2O_3$ is the sum of the masses of the component parts, 159.692 g of $Fe_2O_3$ per 1 mol of $Fe_2O_3$. Percent composition is the mass of a component divided by the mass of the whole times 100. Therefore:

$$\% \text{ Fe in } Fe_2O_3 = \left(\frac{111.694 \text{ g Fe}}{159.692 \text{ g } Fe_2O_3}\right)100$$

$$= 69.9434\% \text{ Fe}$$

**2. 34.59019% Al**
**[5.3]**

First we calculate the mass of 1 mole of gibbsite:

$$? \text{ g Al/mol } Al_2O_3 \cdot 3H_2O = \left(\frac{26.98154 \text{ g Al}}{1 \text{ mol Al}}\right)\left(\frac{2 \text{ mol Al}}{\text{mol } Al_2O_3 \cdot 3H_2O}\right)$$

$$= 53.96308 \text{ Al/1 mol } Al_2O_3 \cdot 2H_2O$$

$$? \text{ g O/mol } Al_2O_3 \cdot 3H_2O = \left(\frac{15.9994 \text{ g O}}{1 \text{ mol O}}\right)\left(\frac{6 \text{ mol O}}{\text{mol } Al_2O_3 \cdot 3H_2O}\right)$$

$$= 95.9964 \text{ g O/1mol } Al_2O_3 \cdot 3H_2O$$

$$? \text{ g H/mol } Al_2O_3 \cdot 3H_2O = \left(\frac{1.0079 \text{ g H}}{1 \text{ mol H}}\right)\left(\frac{6 \text{ mol H}}{\text{mol } Al_2O_3 \cdot 3H_2O}\right)$$

$$= 6.0474 \text{ g H/1 mol } Al_2O_3 \cdot 3H_2O$$

The mass of 1 mole of gibbsite is the sum of the three values just calculated: 53.96308 g + 6.0474 g + 95.9964 g = 156.0069 g. Then we calculate the percentage of aluminum:

$$\% \text{ Al} = \left(\frac{\text{mass Al}}{\text{mass } Al_2O_3 \cdot 3H_2O}\right)100$$

$$= \left(\frac{53.96308 \text{ g}}{156.0069 \text{ g}}\right)100$$

$$= 34.59019\% \text{ Al}$$

**3. $C_5H_5N$**
**[5.2]**

We must determine the ratio of moles of atoms in the molecule. We assume for convenience that there are 100 grams of material. From the percentages given in the problem, there would be 75.92 g of C, 17.71 g of N, and 6.37 g of H in the sample.

$$? \text{ mol C} = 75.92 \text{ g C}\left(\frac{1 \text{ mol C}}{12.011 \text{ g C}}\right)$$

$$= 6.321 \text{ mol C}$$

$$? \text{ mol N} = 17.71 \text{ g N}\left(\frac{1 \text{ mol N}}{14.007 \text{ g N}}\right)$$

$$= 1.264 \text{ mol N}$$

$$? \text{ mol H} = 6.37 \text{ g H}\left(\frac{1 \text{ mol H}}{1.008 \text{ g H}}\right)$$

$$= 6.319 \text{ mol H}$$

The compound has the formula $C_{6.321}N_{1.264}H_{6.3119}$, but this is not an acceptable representation, since atoms combine in whole-number ratios. To reduce any formula to the nearest whole-number ratio, we divide by the smallest number. Thus, the empirical formula is

$$\frac{C_{6.321} \; N_{1.264} \; H_{6.319}}{1.264 \; 1.264 \; 1.264} = C_5NH_5$$

For the same compound the empirical formula is the same as the molecular formula, and the molecule is pyridine, which has the structure

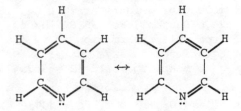

4. empirical formula: $C_{21}H_{27}O_3N$; molecular formula: $C_{42}H_{54}O_6N_2$ [5.2]

We assume for convenience that we have a 100 g sample, and we determine the mole ratios:

$$? \text{ mol C} = 74.0 \text{ g C}\left(\frac{1 \text{ mol C}}{12.01 \text{ g C}}\right)$$

$$= 6.16 \text{ mol C}$$

$$? \text{ mol H} = 7.90 \text{ g H}\left(\frac{1 \text{ mol H}}{1.008 \text{ g H}}\right)$$

$$= 7.84 \text{ mol H}$$

$$? \text{ mol O} = 14.0 \text{ g O}\left(\frac{1 \text{ mol O}}{16.00 \text{ g O}}\right)$$

$$= 0.875 \text{ mol O}$$

$$? \text{ mol N} = 4.10 \text{ g N}\left(\frac{1 \text{ mol N}}{14.01 \text{ g N}}\right)$$

$$= 0.293 \text{ mol N}$$

The nearest whole-number ratio for this compound is $C_{21}H_{27}O_3N$, which has a formula weight of 341 g/mol. The molecular formula is found by determining what multiple the weight of the empirical formula is of the actual molecular weight:

$$\frac{\text{actual molecular weight}}{\text{weight of empirical formula}} = ?$$

$$\frac{682 \text{ g/mol}}{341 \text{ g/mol}} = 2$$

Thus, the molecular formula is twice the empirical formula, and the molecular formula is

$$C_{42}H_{54}O_6N_2$$

5. 3.67 g Fe
   [5.5]

First we make sure that the chemical equation is balanced. The equation as written in the problem is not balanced. The balanced equation

$$Fe_2O_3 + 3CO \rightarrow 2Fe + 3CO_2$$

states that 1 mol $Fe_2O_3$ and 3 mol CO react to give 2 mol Fe and 3 mol $CO_2$. The molecular weights of all compounds can be determined from atomic weights. The molecular weights are: $Fe_2O_3$, 159.692 g/mol; CO, 28.010 g/mol; Fe, 55.847 g/mol; and $CO_2$, 44.010 g/mol. From this information it is obvious that the chemical equation also says that 159.692 g $Fe_2O_3$ and (3 × 28.010) g CO react to yield (2 × 55.847) g Fe and (3 × 44.010) g $CO_2$.

$$Fe_2O_3 + 3CO \rightarrow 2Fe + 3CO_2$$
1 mol $Fe_2O_3$ + 3 mol CO → 2 mol Fe + 3 mol $CO_2$
159.692 g $Fe_2O_3$ + (3 × 28.010) g CO
    → (2 × 55.847) g Fe + (3 × 44.010) g $CO_2$
159.692 g $Fe_2O_3$ + 84.030 g CO
    → 111.694 g Fe + 132.030 g $CO_2$

Remember this type of relationship; such relationships are necessary for solving problems involving chemical reactions. The logic chain can be used to solve the problem:

$$? \text{ g Fe} = 5.24 \text{ g Fe}_2O_3 \left( \frac{(2 \times 55.85) \text{ g Fe}}{159.7 \text{ g Fe}_2O_3} \right)$$

$$= 3.67 \text{ g Fe}$$

As long as we keep track of the units, we can use other logic chains:

$$? \text{ g Fe} = 5.24 \text{ g Fe}_2O_3 \left( \frac{1 \text{ mol Fe}_2O_3}{159.7 \text{ g Fe}_2O_3} \right) \left( \frac{2 \text{ mol Fe}}{1 \text{ mol Fe}_2O_3} \right) \left( \frac{55.85 \text{ g Fe}}{1 \text{ mol Fe}} \right)$$

mol $Fe_2O_3$     mol Fe     g Fe
calculated     calculated     calculated
from molec. → knowing that → using atomic
wt. of $Fe_2O_3$     2 mol Fe     weight of Fe
are obtained
from 1 mol
$Fe_2O_3$

$$= 3.67 \text{ g Fe}$$

A logic chain will always yield the correct answer if we make sure that all unwanted units cancel and all ratios by which we multiply the original data are equal to 1.

6. 55.6%
   [5.5]

From the definition of percent yield and the calculated theoretical yield calculated in problem 5 of this section:

$$\text{percent yield} = \left(\frac{\text{actual yield}}{\text{theoretical yield}}\right)100$$

$$= \left(\frac{2.04 \text{ g}}{3.67 \text{ g}}\right)100$$

$$= 55.6\%$$

7. 3.70 g Fe
   [5.5]

Since all the CO reacts and some $Fe_2O_3$ is left over, the original amount of CO limits the amount of Fe that can be produced:

$$? \text{ g Fe} = 2.79 \text{ g Co} \frac{(2 \times 55.58) \text{ g Fe}}{(3 \times 28.01) \text{ g CO}}$$

$$= 3.70 \text{ g Fe}$$

8. 4.56 g Fe
   [5.5]

First we make sure that all equations are balanced:

$$2C + O_2 \rightarrow 2CO$$

$$2 \text{ mol C} \longrightarrow 2 \text{ mol CO}$$
$$(2 \times 12.011) \text{ g C} \longrightarrow (2 \times 28.010) \text{ g CO}$$

Using the information in the preceding equation and the one for $Fe_2O_3$ reduction (see problem 5 of this section), one can write many logic chains. Some of the following are more efficient than others, but your concern should be to find a method that is logical to you:

$$? \text{ g Fe} = 1.47 \text{ g C} \left(\frac{(2 \times 28.01 \text{ g CO})}{(2 \times 12.01) \text{ g C}}\right)\left(\frac{(2 \times 55.85) \text{ g Fe}}{(3 \times 28.01) \text{ g CO}}\right)$$

$$= 4.56 \text{ g Fe} \qquad or$$

$$? \text{ g Fe} = 1.47 \text{ g C}\left(\frac{1 \text{ mol C}}{12.01 \text{ g C}}\right)\left(\frac{2 \text{ mol Fe}}{3 \text{ mol C}}\right)\left(\frac{55.85 \text{ g Fe}}{1 \text{ mol Fe}}\right)$$

$$= 4.56 \text{ g Fe} \qquad or$$

$$? \text{ g Fe} = 1.47 \text{ g C}\left(\frac{1 \text{ mol C}}{12.01 \text{ g C}}\right)\left(\frac{1 \text{ mol CO}}{1 \text{ mol C}}\right)\left(\frac{2 \text{ mol Fe}}{3 \text{ mol CO}}\right)\left(\frac{55.85 \text{ g}}{1 \text{ mol Fe}}\right)$$

$$= 4.56 \text{ g Fe}$$

9. (a) 2.37 g Fe
   [5.5]

In this problem the quantities of two reagents are specified. Most probably they are not mixed in the exact reaction ratio. Therefore, one of them will not react completely. In order to work the problem we must identify the reagent that reacts completely, since it limits the amount of product that can be obtained:

Method 1:

If we have an electronic calculator, we can solve the problem very quickly by assuming that first one and then the other reagent is limiting. If $Fe_2O_3$ were limiting,

$$? \text{ g Fe} = 4.02 \text{ g } Fe_2O_3\left(\frac{(2 \times 55.85) \text{ g Fe}}{(1 \times 159.69) \text{ g } Fe_2O_3}\right)$$

$$= 2.81 \text{ g Fe}$$

If CO were limiting,

$$? \text{ g Fe} = 1.78 \text{ g CO}\left(\frac{(2 \times 55.85) \text{ g Fe}}{(3 \times 28.01) \text{ g CO}}\right)$$

$$= 2.37 \text{ g Fe}$$

It is obvious that only 2.37 g of Fe could be produced. Since all the CO is consumed to produce this amount of iron, the remaining $Fe_2O_3$ cannot react. The reactants that would yield the least amount of product is the limiting reagent.

Method 2:

First we calculate the number of moles of each reactant:

$$? \text{ mol } Fe_2O_3 = 4.02 \text{ g } Fe_2O_3 \left(\frac{1 \text{ mol } Fe_2O_3}{159.69 \text{ g } Fe_2O_3}\right)$$

$$= 0.02517 \text{ mol } Fe_2O_3$$

$$? \text{ mol} = 1.78 \text{ g CO}\left(\frac{1 \text{ mol CO}}{28.01 \text{ g CO}}\right)$$

$$= 0.06355 \text{ mol CO}$$

The chemical equation says that $Fe_2O_3$ and CO react in a mole ratio of 1 to 3. Therefore, three times the number of moles of $Fe_2O_3$ must be available as CO. In this case there should be $(3 \times 0.0252)$ mol, or 0.0756 mol, of CO available to react completely with the $Fe_2O_3$. This much CO is not available; therefore, all the $Fe_2O_3$ cannot react and CO is the limiting reagent. We use the limiting. reagent to continue the problem:

$$? \text{ g Fe} = 0.06355 \text{ mol CO}\left(\frac{(2 \times 55.85) \text{ g Fe}}{3 \text{ mol CO}}\right) = 2.37 \text{ g Fe}$$

(b) 2.37 g Fe,
    0.00 g CO,
    0.64 g Fe$_2$O$_3$,
    2.80 g CO$_2$,
    [5.5]

After the reaction is complete, 2.37 g of Fe are formed and 0.00 g of CO remains. The mass of CO$_2$ and the mass of residual Fe$_2$O$_3$ need to be calculated:

$$? \text{ g CO}_2 = 0.06355 \text{ mol CO}\left(\frac{(3 \times 44.01) \text{ g CO}}{3 \text{ mol CO}}\right) = 2.80 \text{ g CO}_2$$

The mass of Fe$_2$O$_3$ remaining can be calculated.

$$? \text{ g Fe}_2\text{O}_3 \text{ remaining} = \text{initial g Fe}_2\text{O}_3 - \text{reacted g Fe}_2\text{O}_3$$

$$= 4.02 \text{ g Fe}_2\text{O}_3 - 0.06355 \text{ mol CO}\left(\frac{(1 \times 159.69) \text{ g Fe}_2\text{O}_3}{3 \text{ mol CO}}\right)$$

$$= 4.02 \text{ g Fe}_2\text{O}_3 - 3.38 \text{ g Fe}_2\text{O}_3$$

$$= 0.64 \text{ g Fe}_2\text{O}_3$$

10. 46.1 g C$_2$H$_5$OH
    [5.5]

## II. Thermochemical calculations

Equations [5.9]

C(graphite) + 1/2 O$_2$(g) $\rightarrow$ CO(g)

2Fe(s) + 3/2 O (g) $\rightarrow$ Fe$_2$O$_3$(s)

H$_2$(g) + 1/2 O$_2$(g) $\rightarrow$ H$_2$O(g)

H$_2$(g) + S(s) + 2O$_2$(g) $\rightarrow$ H$_2$SO$_4$(l)

1/2 N$_2$(g) + 3/2 H$_2$(g) $\rightarrow$ NH$_3$(l)

1/2 N$_2$(g) + O$_2$(g) $\rightarrow$ NO$_2$(g)

P(s) + 3/2 Cl$_2$(g) $\rightarrow$ PCl$_3$(g)

P(s) + 1/2 O$_2$(g) + 3/2 Cl$_2$(g) $\rightarrow$ POCl$_3$(g)

S(s) + 3/2 O$_2$(g) $\rightarrow$ SO$_3$(g)

1. $\Delta H$ = -130.5 kJ
   [5.8]

We sum any available equations including the enthalpies such that the final equation and the enthalpy for that equation are obtained. In the final equation SO$_3$(g) is to the left of the reaction arrow, so an equation with known enthalpy change is needed that contains SO$_3$(g) on the left. From the table we choose the equation and standard enthalpy of formation of SO$_3$(g):

S(s) + 3/2 O$_2$(g) $\rightarrow$ SO$_3$(g)     $\Delta H$ = -395.0 kJ

We invert the equation so that SO$_3$(g) is on the left. Inverting the equation changes the sign of the value of $\Delta H$.

SO$_3$(g) $\rightarrow$ S(s) + 3/2 O$_2$(g)     $H$ = +395.0 kJ     (1)

In the final equation $H_2O(l)$ is also needed on the left. From the table we choose the equation and standard enthalpy of formation of $H_2O(l)$, invert the equation, and change the sign of $\Delta H$:

$$H_2O(l) \rightarrow H_2(g) + 1/2\ O_2(g) \qquad \Delta H = +285.8\ kJ \qquad (2)$$

Sulfuric acid is needed on the right in the final equation, so we choose the equation and standard enthalpy of formation of $H_2SO_4(l)$:

$$H_2(g) + S(s) + 2O_2(g) \rightarrow H_2SO_4(l) \quad \Delta H = -811.3\ kJ \quad (3)$$

We sum equations (1), (2), and (3), including the values:

| | |
|---|---|
| $SO_3(g) \rightarrow S(s) + 3/2\ O_2(g)$ | $\Delta H = +395.0\ kJ$ |
| $H_2O(l) \rightarrow H\ (g) + 1/2\ O_2(g)$ | $\Delta H = +285.8\ kJ$ |
| $H_2(g) + S(s) + 2O_2(g) \rightarrow H_2SO_4(l)$ | $\Delta H = -811.3\ kJ$ |
| $SO_3(g) + H_2O(l) \rightarrow H_2SO_4(l)$ | $\Delta H = -130.5\ kJ$ |

To find the enthalpy change of a reaction we need only find appropriate thermochemical data for reactions that, when mathematically combined, give the desired reaction.

*For any reaction, $\Delta H$ equals the sum of the enthalpies of formation of the reactants. Each $\Delta H_f$ must be multiplied by the number of moles that appear before the corresponding molecule in the balanced chemical equation.*

In this problem

$$\Delta H = \Delta H_f \text{ (products)} - \Delta H_f \text{ (reactants)}$$
$$= (-811.3\ kJ) - (-395.0\ kJ - 285.8\ kJ)$$
$$= -130.5\ kJ$$

2. $\Delta H = -136.0\ kJ$
   [5.8]

| | |
|---|---|
| $P(s) + 3/2\ Cl_2(g) + 1/2\ O_2(g) \rightarrow POCl_3(g)$ | $\Delta H = -592.0\ kJ$ |
| $H_2(g) + Cl_2(g) \rightarrow 2HCl(g)$ | $\Delta H = 2(-92.5)\ kJ$ |
| $PCl_5(g) \rightarrow P(s) + 5/2\ Cl_2(g)$ | $\Delta H = +399.2\ kJ$ |
| $H_2O(g) \rightarrow H_2(g) + 1/2\ O_2(g)$ | $\Delta H = +241.8\ kJ$ |
| $PCl_5(g) + H_2O(g) \rightarrow POCl_3(g) + 2HCl(g)$ | $\Delta H = -136.0\ kJ$ |

3. $\Delta H = -147.1\ kJ$    First we balance the equation:
   [5.9]

$$C_5H_{12}(g) + 8O_2(g) \rightarrow 5CO_2(g) + 6H_2O(l) \quad \Delta H = -3536.3\ kJ$$

From the law of Hess we know

$$\Delta H = \Delta H_f \text{ (products)} - \Delta H_f \text{ (reactants)}$$

$$\Delta H_{combustion} = 5\Delta H_f(CO_2) + 6\Delta H_f(H_2O) - \Delta H_f(C_5H_{12}) - 8\Delta H_f(O_2)$$

We rearrange the preceding equation to find the enthalpy of formation of pentane:

$$\Delta H_f(C_5H_{12}) = 5\Delta H_f(CO_2) + 6\Delta H_f(H_2O) - 8\Delta H_f(O_2) - \Delta H_{combustion}$$

Then we substitute values for all known enthalpies:

$$\Delta H_f = 5(-393.7 \text{ kJ}) + 6(-285.8 \text{ kJ}) - 8(0) + 3536.2 \text{ kJ}$$

$$= -147.1 \text{ kJ}$$

4. $\Delta H$
   = -433.0 kJ/mol

We calculate the change in temperature:

$$\Delta T = 26.885°C - 22.815°C = +4.070°C$$

Thus, heat is evolved by the reaction:

$$? \text{ kJ} = (950.0 \text{ g } H_2O + 165.0 \text{ g } H_2O)(4.070°C)(4.184 \text{ J/g°C})$$
$$= 1.899 \times 10^4 \text{ J}$$
$$= 18.99 \text{ kJ}$$

The preceding amount of heat is liberated by the reaction of 12.45 of $P_4O_{10}$. The enthalpy change per mole of $P_4O_{10}$ can be determined as follows:

$$? \text{ kJ/mol} = \left(\frac{18.99 \text{ kJ}}{12.45 \text{ g } P_4O_{10}}\right)\left(\frac{283.9 \text{ g } P_4O_{10}}{1 \text{ mol } P_4O_{10}}\right)$$

$$= -433.0 \text{ kJ}$$

5. $\Delta H$ = -691 kJ
   [5.10]

The lattice energy is the energy associated with the process.

$$K^+(g) + Cl^-(g) \rightarrow KCl(s) \qquad \Delta H_{l.e.} = ?$$

A Born-Haber cycle can be used to determine the value of $\Delta H_{l.e.}$ of KCl. From the law of Hess we know that

$$\Delta H_{l.e.} = \Delta H_f(KCl) - \Delta H_f(K^+) - \Delta H_f(Cl^-)$$

The value of $\Delta H_f(KCl)$ is given, but we must determine $\Delta H_f(K^+)$ and $\Delta H_f(Cl^-)$ from a series of reactions, since $\Delta H_f$ pertains to the formation of a species from elements in standard states. We add the ionization energy of potassium, $\Delta H_{i.p.}$ = 414 kJ, which pertains to the equation

$$K(g) \rightarrow K^+(g) + e^-$$

and the enthalpy of sublimation, $\Delta H_{subl.}$ = 88 kJ, which pertains to the equation

$$K(s) \rightarrow K(g)$$

to obtain a $\Delta H_f$ of $K^+(g)$, $\Delta H_f(K^+)$ = 502 kJ, which pertains to the equation

$$K(s) \rightarrow K^+(g) + e^-$$

Similarly, we add the electron affinity of chlorine, $\Delta H_{e.a.}$ = 368 kJ, which pertains to the equation

$$Cl(g) + e^- \rightarrow Cl^-(g)$$

and half the dissociation energy of $Cl_2$, $1/2\ \Delta H_{diss.}$ = 122 kJ, which pertains to the equation

$$1/2\ Cl_2(g) \rightarrow Cl(g)$$

to obtain $\Delta H_f(Cl^-)$, $\Delta H_f(Cl^-)$ = 246 kJ, which pertains to the equation

$$1/2\ Cl_2(g) + e^- \rightarrow Cl^-(g)$$

To obtain $\Delta H_{1.e.}$ we add the following equations and enthalpies:

$$
\begin{array}{ll}
K(s) + 1/2\ Cl\ (g) \rightarrow KCl(s) & \Delta H_f(KCl) = -435\ kJ \\
K^+(g) + e^- \rightarrow K(s) & -\Delta H_f(K^+) = -502\ kJ \\
\underline{Cl^-(g) \rightarrow 1/2\ Cl\ (g) + e^-} & \underline{-\Delta H_f(Cl^-) =\ 246\ kJ} \\
K^+(g) + Cl^-(g)\quad KCl(s) & \Delta H_{i.e.}(kCl) = -691\ kJ
\end{array}
$$

## SELF-TEST

Complete the test in 20 minutes:

1. Calculate the number of atoms in a 3.20 g sample of copper.

2. In which of the following compounds is the mass percentage of calcium greatest?
   (a) $Ca_3(PO_4)_2$        (c) $CaI_2$
   (b) $CaO$        (d) $CaCO_3$

3. Calculate the enthalpy of formation of $CS_2(l)$ from the following values:

   enthalpy of combustion of $CS_2(l)$ to $CO_2(g)$ and $SO_2(g)$: -1075.7 kJ/mol

   enthalpy of combustion of $C(s)$ to $CO_2(g)$: -393.7 kJ/mol

   enthalpy of combustion of $S(s)$ to $SO_2(g)$: -297.1 kJ/mol

4. A certain organic chemical is shown by analyses to contain by weight 43.90% carbon, 3.05% hydrogen, 9.76% oxygen, and 43.29% chlorine. What is the empirical formula of the compound?

5. The compound $S_4N_3Cl$ can be formed by the reaction

   $$3S_4N_4 + 2S_2Cl_2 \rightarrow 4S_4N_3Cl$$

   How many grams of $S_4N_3Cl$ can be formed with .500 g of $S_4N_4$ and 3.00 g of $S_2Cl_2$?

6. Calculate the enthalpy change of the reaction

   $BaO(s) + CO_2(g) \rightarrow BaCO_3(s)$

   from the following standard enthalpies of formation
   at 298°K: $BaO(s)$, -558.6 kJ/mol; $CO_2(g)$, 393.50
   kJ/mol; and $BaCO_3(s)$, 1216.7 kJ/mol.

# Gases

OBJECTIVES   I. You should be able to demonstrate your knowledge of the
                following terms by defining them, describing them, or giv-
                ing specific examples of them:

                absolute zero [6.3]
                Amonton's Law [6.4]
                atmospheric pressure [6.1]
                Avogadro's principle [6.7, 6.8]
                barometer [6.1]
                Boyle's law [6.2]
                Cannizzaro [6.8]
                Celsius temperature scale [6.3]
                Charles' Law [6.3]
                critical pressure [6.14]
                critical temperature [6.14]
                Dalton's law of partial pressures [6.10]
                density [6.5]

effusion [6.12]
gas [6.1, 6.6]
Gay-Lussac's law of combining volumes [6.7]
Graham's law of effusion [6.12]
ideal gas [6.5]
Joule-Thomson effect [6.14]
Kelvin (also called absolute) temperature scale [6.3]
kinetic energy [6.6]
manometer [6.1]
pressure [6.1]
real gas [6.13]
standard temperature and pressure, abbreviated STP [6.5]
STP molar volume [6.7]
torr [6.1]
vapor pressure [6.10]
van der Waals equation [6.13]

II. If all but one of the variables (pressure, $P$, volume, $V$, number of moles, $n$, and temperature, $T$) of the ideal gas law are given, you should be able to calculate the unknown variable.

III. If a gas under a defined set of conditions is changed to a new set of conditions, you should be able to use Boyle's, Charles's, and Amonton's laws in calculations to determine the complete new set of conditions.

IV. You should understand vapor pressure and partial pressures and be able to perform calculations involving them.

V. You should be able to calculate relative rates of gaseous fusion.

VI. You should understand Gay Lussac's law of combining volumes and be able to perform calculations involving chemical reactions of gases.

## EXERCISES

I. Answer each of the following with *true* or *false*. If a statement is false, correct it.

_____    1. One atmosphere equals 760 torr.

_____    2. Boyle's law states that the volume of a gas varies inversely with the pressure under which it is measured.

_____    3. Charles's law states that the volume of a gas varies inversely with the pressure under which it is measured.

_____    4. STP stands for standard temperature, 273°K, and pressure, 1 atmosphere.

_____    5. The value of $R$ is 0.08206 liter atm $K^{-1}$ $mol^{-1}$.

_____    6. $PV = nRT$.

_____    7. Temperatures in °C are changed to the Kelvin scale by adding 100: $T = t + 100$.

_____    8. A mole of ideal gas contains $6.022 \times 10^{23}$ molecules and occupies 22.414 liters at STP.

_____    9. In a mixture of gases the total pressure is the sum of the partial pressures of the components.

_____    10. According to Graham's law of effusion, gases with larger molecular weights effuse through small openings more rapidly than gases with smaller molecular weights.

_____    11. The van der Waals equation accounts for the fact that real gas molecules have no volume and exert no attractive forces.

_____    12. Below the critical temperature it is impossible to liquefy a gas regardless of pressure.

II. Complete the following statements with one of the following:

    increases
    decreases
    remains the same

1. If the temperature of a gas is increased and the pressure on the system is unchanged, the volume of the gas _____.

2. If a gas is enclosed in a rigid container and the container is heated, the pressure exerted by the gas _____.

3. A vessel contains 2.5 mol of oxygen. If an additional 2.5 mol of oxygen is added to the vessel, the pressure _____.

4. A gas is allowed to expand from 1 liter to 22.4 liters. The number of moles of gas _____.

5. As a gas is heated, the average kinetic energy of the molecules _____.

6. As a gas is compressed without changing the temperature, the average kinetic energy of the molecules _____.

7. As a gas is compressed at constant temperature, the mean free path of a molecule _____.

8. As the volume of a gas is increased at constant temperature, the pressure _____.

9. A cylinder contains oxygen at a pressure of 1500 pounds per square inch, and some gas is released from it. The pressure _____. The number of moles of gas in the cylinder _____. If there is no temperature change, the average kinetic energy of the molecules in the cylinder _____ and the mean free path _____.

10. An evacuated 1-liter flask is opened at sea level. If there is no temperature change, the number of moles of gas in the flask _____. The flask is sealed and carried to a mountain top. During transport the pressure of the contained gas _____. The flask is again opened. The number of moles of gas in the flask _____ and the pressure _____.

11. A weather balloon is released from a station in Texas. As the balloon rises, its size _____ due to decreased atmospheric pressure.

12. A diver carries a tank that could supply a 30-minute oxygen supply at sea level. The effective oxygen supply _____ as the diver explores at a depth of 50 feet if temperature is constant.

13. The combustion of hydrogen is represented by

    $$2H_2(g) + O_2(g) \rightarrow 2H_2O(g)$$

    If the temperature and volume do not change, the pressure _____ as the reaction proceeds in a closed container. If an appreciable temperature increase occurs during the reaction, the final pressure _____ due to the temperature change.

14. A gas in a 1-liter container is heated from 0°C to 100°C. Simultaneously the volume of the container increases to 2 liters. The pressure _____ due to the temperature change the _____ due to the volume change.

15. Octane is burned at constant pressure in an expandable vessel:

$$2C_8H_{18}(g) + 25O_2(g) \rightarrow 16CO_2(g) + 18H_2O(g)$$

As the reaction proceeds, the number of moles of gas _____ and the volume of gas _____.

16. A naturally occurring mixture of $^{35}Cl$ and $^{37}Cl$ is enclosed in a cylinder. As a small leak allows gas to escape very slowly, a piston maintains the trapped sample at STP. The number of molecules in the cylinder _____. The density of the trapped gas _____. The molecular weight of the trapped gas _____.

17. A mole of helium at STP is enclosed in a rigid container. Argon is gradually added to the container. The pressure _____. The number of moles of helium _____. The partial pressure of helium _____. The partial pressure of argon _____.

18. A rigid 1-liter vessel contains equimolar concentrations of He and Ar. As the temperature is increased, the partial pressure of He, $p_{He}$, _____ and the partial pressure of Ar, $p_{Ar}$, _____. The total pressure _____.

19. Gas is gradually escaping from a container. The number of moles of gas, $n$, _____. The volume and pressure are maintained constant. The temperature of the gas _____.

III. Work the following problems:

1. A vessel containing argon at 0.10 atm is heated from 0°C to 100°C. If the volume does not change, what is the final pressure?

2. A 75-ml sample of gas is heated at constant pressure from 0°C to 33°C. What is the volume at 33°C?

3. What volume will an ideal gas occupy at absolute zero?

4. A gas in a 1.00-liter container is allowed to expand to a volume of 5.00 liters. If the initial pressure is 748 torr, what is the final pressure if the temperature does not change?

5. A 50-ml gas sample is cooled from 100°C to 12°C while the pressure is held constant. What is the final volume?

6. What volume will 16 g of oxygen gas occupy at STP?

7. What volume in ml will 0.500 mol of an ideal gas occupy at STP?

8. What is the density, in g/ml, of oxygen at STP?

9. How many moles of gas at 1.00 atm and 25°C are contained in a 5.00-liter vessel?

10. A balloon containing 1.00 liter of He at STP is purchased in an air-conditioned store, in which the temperature is 25°C, and carried outside where it heats to 45°C. If there is no pressure change, what is the final volume?

11. If the pressure on the balloon described in problem 10 of this section does change from 745 torr to 757 torr during the temperature change, what is the final volume?

12. A gas sample is heated from -10°C to 87°C, and the volume is increased from 1.00 liter to 3.30 liters. If the initial pressure is 0.750 atm, what is the final pressure?

13. A 1.0-liter flask contains 3.5 mol of gas at 27°C. What is the pressure of the gas?

14. An evacuated 1.00-liter flask weighs 104.35 g. A quantity of nitrogen gas is added to the flask. After the nitrogen is added, the flask and contents weigh 105.68 g, and the pressure exerted by the gas is 2.00 atm. What is the temperature inside the flask?

15. When magnesium metal is burned in air, solid MgO forms:

$$2Mg(s) + O_2(g) \rightarrow 2MgO(s)$$

What volume of oxygen at STP is needed to react with 0.500 g of Mg?

16. The complete combustion of octane yields carbon dioxide and water:

$$2C_8H_{18}(g) + 25O_2(g) \rightarrow 16CO_2(g) + 18H_2O(g)$$

What volume of gas is produced from the complete combustion of 0.670 g of octane if the temperature is 400°C and the pressure is 1.2 atm?

17. Hydrogen gas can be generated by the action of hydrochloric acid on magnesium:

$$Mg(s) + 2HCl(aq) \rightarrow Mg^{+2}(aq) + 2Cl^-(aq) + H_2(g)$$

If all the hydrogen from the reaction of 2.00 g of Mg is collected over water at 1.00 atm and 25°C, what volume will the dried gas occupy at STP?

18. Traces of oxygen can be removed from streams of gas by reaction with heated copper:

$$2Cu(s) + O_2(g) \rightarrow 2CuO(s)$$

If the original copper weighs 0.3750 g and the oxidized copper weighs 0.3800 g, what volume of $O_2$ at STP was removed from the gas stream?

19. In a normal breath 2.00 liters of gas can be inhaled. How many moles of gas at STP are in this volume? If 21 percent (by volume) of the inhaled gas molecules is oxygen, how many grams of oxygen are in this volume?

20. The atmospheric pressure on Mt. Everest is 0.330 atm with a temperature of -10°C? How many grams of oxygen are inhaled in a 2.00-liter breath?

21. What is the total percentage decrease in oxygen intake per breath when a person goes from sea level to the mountain top? (See problems 19 and 20 of this section.)

22. If pure oxygen is inhaled from an oxygen cylinder on a mountain top at 0.330 atm and -10°C, how many grams of oxygen would be inhaled per 2.00-liter breath?

23. One mole of hemoglobin, which has a weight of 66,280 g, can bind 4 moles of oxygen gas. What volume of $O_2$ at STP can 100 ml of blood with a hemoglobin concentration of 160 g/liter carry?

24. Sulfur dioxide concentrations in the atmosphere can easily reach 0.20 ppm, which is 0.20 ml of $SO_2$ per $10^6$ ml of air. How many molecules of $SO_2$ at STP are inhaled in a 2.00-liter breath?

25. About $1.5 \times 10^8$ metric tons (1 metric ton = 1000 kg) of CO are released into the atmosphere each year. What is the volume of this quantity of CO at STP?

26. At STP 0.66 mol of $H_2$ and 0.33 mol of $O_2$ are mixed in an expandable container. After ignition the product of the reaction reaches a temperature of 1300°C with a pressure of 800 torr. What is the volume of the reaction product?

$$2H_2(g) + O_2(g) \rightarrow 2H_2O(g)$$

27. At STP 0.66 mol of $H_2$ and 0.66 mol of $O_2$ are mixed in an expandable container. After ignition the remaining gases reach a temperature of 1300°C and a pressure of 800 torr. What is the total volume of the remaining gases?

28. Atmospheric air is a gaseous mixture of water vapor, oxygen, carbon dioxide, nitrogen, and traces of other species. What is the mole fraction of nitrogen gas at at STP if $p_{O_2}$ = 159 torr, $p_{CO_2}$ = 0.23 torr, and $p_{H_2O}$ = 23.8 torr?

29. A gas, X, diffuses 3.1 times faster than fluorine gas. What is the molecular weight of the gas X?

30. A storage tank at JFK Space Center can contain 900,000 gallons, which is approximatly 3.4 million liters, of liquid hydrogen, which has a density of 0.070 g/cc at −253°C. What volume in liters would this hydrogen occupy at STP if it were evaporated?

31. W. W. Ruby estimated the total mass of atmospheric oxygen to be $15 \times 10^{20}$ g. If all this oxygen were at STP, what volume in liters would it occupy?

32. Uranium 235 and uranium 238 are separated by the effusion difference of the hexafluorides $^{235}UF_6$ and $^{238}UF_6$. What is the ratio of the effusion rates of the hexafluorides?

33. What is the density of phosgene gas, $COCl_2$, at STP?

34. A 100-ml flask contains 0.162 g of an unknown gas at 760 torr and 100°C. What is the molecular weight of the gas?

35. Acetylene is formed by the reaction of water and calcium carbide:

$$CaC_2(s) + 2H_2O(l) \rightarrow C_2H_2(g) + Ca(OH)_2(s)$$

What volume of acetylene can be produced at STP from 4.00 g of $CaC_2$ and 2.75 g of water?

36. Acetylene is burned to form water and carbon dioxide:

$$2C_2H_2(g) + 5O_2(g) \rightarrow 2H_2O(g) + 4CO_2(g)$$

Answer the following:
(a) What volume of $CO_2$ at STP is formed from 3.00 g of $C_2H_2$?
(b) What volume of $CO_2$ at STP is formed from 2.00 liters of $C_2H_2$ at STP?

37. The van der Waals constants for ammonia are given in Table 6.4 of your text. Use this data to determine the pressure at which 1 mole of $NH_3$ will occupy 22.4 liters at 25°C.

38. Use the date in Table 6.4 of your text to determine which of the gases listed has the strongest inter-molecular interactions and which has the largest molecular volume.

39. One (1.0) ml of liquid water at 25°C, which has den-sity of 1.00 g/ml, is placed in an evacuated 10-liter vessel, which is also at 25°C. What is the pressure inside the vessel? What is the mass of water in the gaseous state?

40. One (1.00) ml of liquid water at 25°C is added to dry helium that is contained in a 10.0-liter cylinder at 25°C and 100 atm. What is the partial pressure of water vapor in the vessel? What is the partial pres-sure of He? What is the total pressure?

41. At what temperature will all the water in problem 39 be volatilized and exert a pressure of 1.0 atm? The density of water at 25°C is 1.0 g/ml.

42. One (1.0) liter of a gas collected over water at STP weighs 1.135 g before drying. What is the molecular weight of the unknown gas?

43. A gas mixture is known to contain only helium and nitrogen. What are the partial pressures of each gas if the density of the mixture is 0.475 g/liter at STP?

44. What would be the total volume of the atmosphere at STP if the atmospheric concentration of $O_2$ were 21 percent by volume? (See problem 31 of this section.)

45. Would the actual volume of the earth's atmosphere be larger or smaller than that estimated in problem 44 of this section? Why?

46. In the atmosphere upon irradiation with ultraviolet light, oxygen is converted to ozone, $O_3$:

$$3O_2 \xrightarrow{h\nu} 2O_3$$

The symbol $h\nu$ above the arrow in the equation for the reaction indicates the quantum of radiation necessary to cause the reaction to proceed. This same reaction can be carried out under controlled laboratory conditions. If 1.47 liters of $O_2$ in a sealed container at 1.01 atm is irradiated until 5.24

percent of the oxygen reacts, what is the final pressure inside the flask? Assume that the temperature also rises from 25°C to 54°C during the reaction.

## ANSWERS TO EXERCISES

### I. Properties of gases

1. True [6.1]

2. True [6.2]

3. False [6.3]    Volume varies directly with temperature in K.

4. True [6.5]

5. True [6.5]

6. True [6.5]

7. False [6.3]    $T = t + 273$

8. True [5.1, 6.7]

9. True [6.10]

10. False [6.12]    A molecule with a smaller molecular weight diffuses more quickly.

11. False [6.13]    In the van der Waals equation $a$ corrects for intermolecular interactions and $b$ corrects for molecular volume.

12. False [6.14]    Above the critical temperature liquefaction cannot occur.

### II. Behavior of gases

1. increases [6.3]

2. increases [6.3]

3. increases [6.5]

4. remains the same [6.5]

5. increases [6.6]

6. remains the same [6.6]

7. decreases [6.6]

8. decreases [6.2]

9. decreases [6.5]    Gas escapes.
   decreases [6.5]
   remains the same [6.6]
   increases [6.6]    Molecules are less tightly packed.

10. increases [6.5]    Gas enters the flask until the pressure inside the tank
                        equals the pressure outside.

    remains the        The volume, number of moles, and temperature remain con-
    same [6.5]         stant.
    decreases [6.5]    Gas escapes.
    decreases [6.5]

11. increases [6.2]

12. decreases [6.5]    More gas will remain in the tank at the higher pressures
                        encountered beneath sea level. Since $PV = nRT$, and $V$, $R$,
                        and $T$ are constant in this problem, it is necessary for
                        $n$ to be larger if $P$ is larger. $P$ is larger under the
                        sea, so $n$, the number of moles in the tank, must also
                        be larger.

13. decreases [6.8]    Three moles of reactants produce only 2 moles of product.
    increases

14. increases [6.5]
    decreases [6.5]

15. increases [6.5]
    increases [6.5]

16. decreases [6.5]
    increases [6.12]   The $^{35}Cl$ isotope effuses more quickly than $^{37}Cl$; there-
                        fore, the remaining gas is enriched in the heavier iso-
                        tope.

    increases [2.9]    Atomic weights are weighted averages of isotopic masses.

17. increases [6.10]
    remains the
    same [6.5]
    remains the
    same [6.10]
    increases [6.5]

18. increases
    [6.5, 6.10]
    increases
    [6.5, 6.10]
    increases [6.6]

19. decreases [6.5]
    increases [6.5]

III. Gas law calculations

**1. 0.14 atm**
**[1.5, 6.4]**

An increase in temperature results in an increase in pressure. In the gas laws temperatures must be expressed in K, whereas pressures may be expressed in any pressure units. (See Example 6.3 of your text.) Thus,

$$? \text{ atm} = 0.10 \text{ atm} \left( \begin{array}{c} \text{temperature} \\ \text{correction} \\ \text{factor} \end{array} \right)$$

Notice that the ratio of temperatures must be larger than 1 to reflect an increase in pressure:

$$? \text{ atm} = 0.10 \text{ atm} \left( \frac{373 \text{ K}}{273 \text{ K}} \right) = 0.14 \text{ atm}$$

Note that only two significant figures are reported since 0.10 atm has two significant figures. You may find it convenient to tabulate information for use in Boyle's and Charles's laws in the following way. Tabulations such as this should be helpful if you have a tendency to invert correction factors.

|       | Initial Conditions | Final Conditions |
|-------|--------------------|------------------|
| $T$   | (0 + 273) K        | (100 + 273) K    |
| $P$   | 0.10 atm           | ?                |
| $V$   | constant           | constant         |

Since the temperature increases and the volume is constant, the pressure must increase; therefore, the temperature correction factor must be larger than 1.

$$? \text{ atm} = 0.10 \text{ atm} \left( \frac{373 \text{ K}}{273 \text{ K}} \right) = 0.14 \text{ atm}$$

**2. 84 [6.3]**

An increase in temperature is accompanied by an increase in volume. In the gas laws volume may be expressed in any volume units, whereas temperature must be expressed in K. Thus, we find

$$? \text{ ml} = 75 \text{ ml} \left( \frac{306 \text{ K}}{273 \text{ K}} \right) = 84 \text{ ml}$$

**3. 0 [6.5]**

An ideal gas has no molecular volume and compresses to zero volume at absolute zero.

**4. 150 torr [6.3]**

An increase in volume is accompanied by a decrease in pressure. (See Example 6.2 of your text.) Thus, we find

$$? \text{ atm} = 748 \text{ torr} \left( \frac{1.00 \text{ liter}}{5.00 \text{ liter}} \right) = 150 \text{ torr}$$

5. 38 ml [6.3]     A decrease in temperature is accompanied by a decrease in volume. Thus, we find

$$? \text{ ml} = 50 \text{ ml}\left(\frac{285 \text{ K}}{373 \text{ K}}\right) = 38 \text{ ml}$$

According to Charles's law, the volume of an ideal gas varies directly with absolute temperature. (See Example 6.3 of your text.)

6. 11 liters $O_2$     First we calculate the number of moles of oxygen gas.
   [6.5, 6.7]     Oxygen is diatomic; i.e., it exists as $O_2$ molecules, and the molecular weight is 32.0.

$$? \text{ mol } O_2 = \frac{16 \text{ g } O_2}{32.0 \text{ g } O_2/1 \text{ mol } O_2} = 0.500 \text{ mol } O_2$$

Each mole of ideal gas occupies 22.4 liters at STP, i.e., 273 K and 1 atm. Therefore,

$$? \text{ liter } O_2 = 0.500 \text{ mol } O_2\left(\frac{22.4 \text{ liters } O_2}{1 \text{ mol } O_2}\right)$$

$$= 11 \text{ liters } O_2$$

We can also directly use the ideal gas law:

$$PV = nRT$$

$$V = \frac{nRT}{P}$$

$$= \frac{(0.500 \text{ mol})(0.0821 \text{ liter atm K}^{-1} \text{ mol}^{-1})(273 \text{ K})}{1 \text{ atm}}$$

$$= 11 \text{ liters}$$

7. $1.12 \times 10^4$ ml     A mole of an ideal gas occupies 22.4 liters at STP,
   [6.7]     and each liter contains 1000 ml. Therefore.

$$? \text{ ml} = 0.500 \text{ mol}\left(\frac{22.4 \text{ liters}}{1 \text{ mol}}\right)\left(\frac{1000 \text{ ml}}{1 \text{ liter}}\right)$$

$$= 1.12 \times 10^4 \text{ ml}$$

Note that all units cancel except ml. It is imperative to include proper units in all calculations.

8. $1.43 \times 10^{-3}$ g/ml     At STP a mole of oxygen occupies 22.4 liters. Each mole
   [6.7]     weighs 32.0 g. Therefore,

$$? \text{ g/ml} = \left(\frac{1 \text{ mol } O_2}{22.4 \text{ liters } O_2}\right)\left(\frac{32.0 \text{ g } O_2}{1 \text{ mol } O_2}\right)\left(\frac{1 \text{ liter } O_2}{1000 \text{ ml } O_2}\right)$$

$$= 1.43 \times 10^{-3} \text{ g/ml}$$

9. 0.204 mol
   [6.5]

We use the ideal gas law:

$$PV = nRT$$

$$n = \frac{(1.00 \text{ atm})(5.00 \text{ liters})}{(0.08206 \text{ liter atm K}^{-1} \text{ mol}^{-1})(298 \text{ K})}$$

$$= 0.204 \text{ mol}$$

If 0.08206 liter atm $K^{-1}$ $mol^{-1}$ is the value used for $R$, volume must be expressed in liters, temperature in K, and pressure in atmospheres.

10. 1.07 liters
    [6.3]

According to Charles's law, an increase in temperature is accompanied by an increase in volume. Therefore,

$$? \text{ liters} = (1.00 \text{ liter})\left(\frac{318 \text{ K}}{298 \text{ K}}\right) = 1.07 \text{ liter}$$

11. 1.05 liters
    [6.3]

Charles's law predicts that an increase in temperature is accompanied by an increase in volume, and Boyle's law predicts that an increase in pressure is accompanied by a decrease in volume. Combining laws, we find

$$? \text{ liters} = (1.00 \text{ liter})\left(\frac{318 \text{ K}}{298 \text{ K}}\right)\left(\frac{745 \text{ torr}}{757 \text{ torr}}\right)$$

$$= 1.05 \text{ liters}$$

Note that the temperature correction factor must be larger than 1 to reflect an increase in volume. The pressure correction factor must be smaller than 1 to account for a decrease in volume due to an increase in pressure. The pressure change does not have a signifi cant effect in this problem; however, this is not the normal situation.

12. 0.311 atm
    [6.2, 6.3,
    6.4]

An increase in temperature causes an increase in pressure, and an increase in volume causes a decrease in pressure. Therefore,

$$? \text{ atm} = 0.750 \text{ atm}\left(\frac{360 \text{ K}}{263 \text{ K}}\right)\left(\frac{1.00 \text{ liter}}{3.30 \text{ liters}}\right)$$

$$= 0.311 \text{ atm}$$

If you did not solve this problem correctly the first time, tabulate the data as shown in problem 1 of this section:

|   | Initial Conditions | Final Conditions |
|---|---|---|
| $T$ | (-10 + 273) K | (87 + 273) K |
| $P$ | 0.750 atm | ? |
| $V$ | 1.00 liter | 3.30 liters |

Use the preceding logic to solve the problem.

13. 86 atm [6.5]    We use the ideal gas law:

$$PV = nRT$$

$$P = \frac{nRT}{V}$$

$$= \frac{(3.5 \text{ mol})(0.0821 \text{ liter atm K}^{-1} \text{ mol}^{-1})(300 \text{ K})}{1.0 \text{ liter}}$$

$$= 86 \text{ atm}$$

If we wish the pressure to be expressed in torr, we convert atmosphere to torr:

$$? \text{ torr} = 86 \text{ atom}\left(\frac{760 \text{ torr}}{1 \text{ atm}}\right) = 6.5 \times 10^4 \text{ torr}$$

14. 513 K, or 240°C [6.5]    If a problem concerns an ideal gas and temperature, pressure, and volume are not changed, we can use $PV = nRT$. In this problem we wish to calculate temperature:

$$T = \frac{PV}{nR}$$

Since $P$, $V$, and $R$ are known, we need to compute the value of $n$:

$$? \text{ mol N} = \frac{(105.68 - 104.35) \text{ g N}_2}{(28.02 \text{ g N}_2/1 \text{ mol N}_2)} = 0.04747 \text{ mol N}_2$$

Substituting values into the ideal gas law, we find

$$T = \frac{(2.00 \text{ atm})(1.00 \text{ liter})}{(0.04747 \text{ mol})(0.08206 \text{ liter atm K}^{-1} \text{ mol}^{-1})}$$

$$= 513 \text{ K}$$

and

$$T = (513 - 273)(1°C) = 240°C$$

15. 0.230 liter $O_2$ [6.9]    The balanced equation shows that 1 mol of $O_2$, which occupies 22.4 liters at STP, combines with 2 mol of Mg, $(2 \times 24.31)$ g Mg. This relationship can be used to compute the volume of $O_2$ gas:

$$? \text{ liters } O_2 = 0.500 \text{ g Mg}\left(\frac{22.4 \text{ liters } O_2}{(2 \times 24.31) \text{ g Mg}}\right)$$

$$= 0.230 \text{ liter } O_2$$

See example 6.14 of your text.

16. 4.6 liters [6.8]    Compute the volume of octane at STP:

$$= \left(\frac{0.670 \text{ g } C_8H_{18}}{(114.2 \text{ g } C_8H_{18}/1 \text{ mol})}\right)\left(\frac{22.4 \text{ liters } C_8H_{18}}{1 \text{ mol}}\right)$$

$$= 0.131 \text{ liter}$$

The balanced equation shows that 16/2, or 8, times the volume of octane is the volume of $CO_2$ produced, and 18/2, or 9, times the volume of octane is the volume of water vapor produced. Therefore, the total volume of gas produced at STP would be

$$V_{total} = V_{CO_2} + V_{H_2O}$$

$$= (8 \times 0.131) \text{ liter} + (9 \times 0.131) \text{ liter}$$

$$= 2.23 \text{ liter}$$

Changing from STP to the actual conditions:

$$? \text{ liters} = 2.23 \text{ liters} \left(\frac{(673 \text{ K})(1.0 \text{ atm})}{(273 \text{ K})(1.2 \text{ atm})}\right)$$

$$= 4.6 \text{ liters gas}$$

17. 1.84 liters $H_2$
    [6.8]

Since the water vapor is removed before the volume is measured, its effect need not be considered. Therefore,

$$? \text{ liters } H_2 = 2.00 \text{ g Mg } \frac{22.4 \text{ liters } H_2}{24.30 \text{ g Mg}}$$

$$= 1.84 \text{ liters } H_2$$

18. 0.0035 liter, or 3.5 ml, $O_2$

We calculate the mass of oxygen:

$$? \text{ g} = 0.3800 \text{ g} - 0.3750 \text{ g} = 0.0050 \text{ g}$$

We calculate the volume occupied by this quantity of oxygen at STP

$$? \text{ liters } O_2 = \left(\frac{0.0050 \text{ g } O_2}{(32.0 \text{ g } O_2/1 \text{ mol } O_2)}\right)\left(\frac{22.4 \text{ liters } O_2}{1 \text{ mol } O_2}\right)$$

$$= 0.0035 \text{ liter}$$

or

$$? \text{ ml } O_2 = 0.0035 \text{ liter } (1000 \text{ ml}/1 \text{ liter}) = 3.5 \text{ ml}$$

19. 0.60 g $O_2$
    [6.5, 6.7]

First we calculate the number of moles of gas at STP:

$$PV = nRT$$

$$n = \frac{PV}{RT}$$

$$= \frac{(1.00 \text{ atm})(2.00 \text{ liters})}{(0.08206 \text{ liter atm } K^{-1} \text{ mol}^{-1})(273 \text{ K})}$$

$$= 8.92 \times 10^{-2} \text{ mol}$$

Alternatively, since 1 mol of ideal gas at STP occupies 22.4 liters,

$$? \text{ mol gas} = 2.00 \text{ liters} \left( \frac{1 \text{ mol gas}}{22.4 \text{ liters gas}} \right)$$

$$= 8.93 \times 10^{-2} \text{ mol gas}$$

Twenty-one percent of this quantity is $O_2$. Therefore,

$$? \text{ g } O_2 = 8.93 \times 10^{-2} \text{ mol gas} \left( \frac{0.21 \text{ mol } O_2}{1 \text{ mol gas}} \right) \left( \frac{32.0 \text{ g } O_2}{1 \text{ mol } O_2} \right)$$

$$= 0.60 \text{ g } O_2$$

**20.** 0.20 g $O_2$
  [6.5]

On the top of Mt. Everest,

$$n = \frac{(0.330 \text{ atm})(2.00 \text{ liters})}{(0.0821 \text{ liter atm K}^{-1} \text{ mol}^{-1})(263 \text{ K})}$$

$$= 3.06 \times 10^{-2} \text{ mol}$$

and

$$? \text{ g } O_2 = 3.06 \times 10^{-2} \text{ mol gas} \left( \frac{0.21 \text{ mol } O_2}{1 \text{ mol gas}} \right) \left( \frac{32.0 \text{ g } O_2}{1 \text{ mol } O_2} \right)$$

$$= 0.20 \text{ g } O_2$$

**21.** 67 percent

The total percent *decrease* is calculated by dividing the difference of the two values by the higher value and then multiplying by 100:

$$? \text{ \% decrease} = \left( \frac{(0.60 - 0.20) \text{ g } O_2}{0.60 \text{ g } O_2} \right) (100) = 67\%$$

If the problem had asked for the total percent *increase* in oxygen intake per breath when a person goes from the mountain top to sea level, 0.20 g of $O_2$ would be in the denominator. In percentage calculations the correct value or reference value is always in the denominator.

**22.** 0.979 g $O_2$

We use the ideal gas law to calculate $n$:

$$PV = nRT$$

$$n = \frac{PV}{RT}$$

$$= \frac{(0.330 \text{ atm})(2.00 \text{ liters})}{(0.08206 \text{ liter atm K}^{-1} \text{ mol}^{-1})(263 \text{ K})}$$

$$= 3.06 \times 10^{-2} \text{ mol}$$

Then we calculate the grams of $O_2$:

$$? \text{ g } O_2 = 3.06 \times 10^{-2} \text{ mol } O_2 \ (32.0 \text{ g } O_2/1 \text{ mol } O_2)$$

$$= 0.979 \text{ g } O_2$$

Note that more oxygen is inhaled per 2.00-liter breath when a person breathes from a cylinder of oxygen on Mt. Everest than when he or she breathes normally at sea level.

23. 21.6 mol $O_2$
    [6.8]

This problem describes a balanced chemical equation. Note that 1 mole of hemoglobin combines with 4 moles of $O_2$. Therefore, if we let Hb stand for hemoglobin, the equation is

$$Hb + 4O_2 \rightarrow Hb \cdot 4O_2$$

in which $Hb \cdot 4O_2$ is the product. The problem can be solved simply be determining the number of grams of Hb available:

$$? \text{ liters } O_2 = 16.0 \text{ g Hb } \left( \frac{(4 \times 22.4) \text{ liters } O_2}{66,280 \text{ g Hb}} \right)$$

$$= 0.0216 \text{ liter } O_2$$

or

$$? \text{ ml } O_2 = 0.0216 \text{ liter } O_2 \left( \frac{1000 \text{ ml } O_2}{1 \text{ liter } O_2} \right)$$

$$= 21.6 \text{ ml } O_2$$

24. $1.1 \times 10^{16}$
    molecules $SO_2$
    [6.7]

First we calculate the volume of $SO_2$:

$$? \text{ liters } SO_2 = 2.00 \text{ liters air } \left( \frac{0.20 \text{ liter } SO_2}{1 \times 10^6 \text{ liters air}} \right)$$

$$= 4.00 \times 10^{-7} \text{ liter } SO_2$$

Then we calculate the number of moles of $SO_2$ from the volume of $SO_2$, and finally we calculate the number of molecules of $SO_2$ from the number of moles of $SO_2$. Combining these two calculations, we find

$$? \text{ molecules } SO_2$$

$$= 4.00 \times 10^{-7} \text{ liters } SO_2 \left( \frac{6.02 \times 10^{23} \text{ molecules } SO_2}{22.4 \text{ liters } SO_2} \right)$$

$$= 1.1 \times 10^{16} \text{ molecules } SO_2$$

Can you calculate the mass of $SO_2$?

25. $1.2 \times 10^{14}$
    liters CO
    [6.7]

We calculate the volume at STP:

$$? \text{ liters CO}$$

$$= 1.5 \times 10^8 \text{ metric tons CO } \left( \frac{10^6 \text{ g}}{1 \text{ metric ton}} \right) \left( \frac{22.4 \text{ liters CO}}{28.0 \text{ g CO}} \right)$$

$$= 1.2 \times 10^{14} \text{ liters CO}$$

The factors used are

$$10^3 \text{ kg} = 1 \text{ metric ton}$$

$$10^3 \text{ g} = 1 \text{ kg}$$

22.4 liters CO = 1 mol CO = 28.0 g CO

See Section 5.2 of your text.

26. 81 liters $H_2O$     The oxygen and hydrogen are mixed in the proper molar
    [6.9]                ratio for complete reaction, 2 mol $H_2$ to 1 mol $O_2$.
                         Therefore, either the number of moles of $O_2$ or $H_2$ can
                         be used to begin the calculation. First we calculate
                         the volume of water vapor at STP:

$$? \text{ liters } H_2O = 0.33 \text{ mol } O_2 \left( \frac{(2 \times 22.4) \text{ liters } H_2O}{1 \text{ mol } O_2} \right) = 14.8 \text{ liters } H_2O$$

or

$$? \text{ liters } H_2O = 0.66 \text{ mol } H_2 \left( \frac{(2 \times 22.4) \text{ liters } H_2O}{2 \text{ mol } H_2} \right) = 14.8 \text{ liters } H_2O$$

Then we adjust the volume to the actual temperature and
pressure conditions. An increase in temperature is accompanied by an increase in volume, and an increase in pressure is accompanied by a decrease in volume. Thus,

$$? \text{ liters } H_2O = 14.8 \text{ liters } H_2O \left( \frac{1573 \text{ K}}{273 \text{ K}} \right) \left( \frac{760 \text{ torr}}{800 \text{ torr}} \right)$$

$$= 81 \text{ liters } H_2O$$

27. $1.2 \times 10^2$     It takes only 0.33 mol $O_2$ to react completely with all of
                         the hydrogen. The excess $O_2$, 0.33 mol $O_2$, will remain in
                         the container. The volume occupied by $H_2O(g)$ is 81 liters
                         (see problem 25 of this section). The volume occupied by
                         the excess $O_2$ is

$$V = \frac{(0.33 \text{ mol})(0.0821 \text{ liter atm K}^{-1} \text{ mol}^{-1})(1573 \text{ K})}{800 \text{ torr } \frac{1 \text{ atm}}{760 \text{ torr}}} = 40 \text{ liters}$$

The total volume of all gases remaining after ignition is

$$V_{total} = 81 \text{ liters } H_2O + 40 \text{ liters } O_2$$

$$= 121 \text{ liters gas}$$

Alternatively, we could reason that 0.66 mol $H_2$ reacts
to form 0.66 mol $H_2O$ and 0.33 mol $O_2$ remains unreacted.
A total of 0.99 mol gas, 0.66 mol $H_2O$, and 0.33 mol $O_2$
remains after the reaction. Therefore,

$$V = \frac{(0.99 \text{ mol})(0.0821 \text{ liter atm K}^{-1} \text{ mol}^{-1})(1573 \text{ K})}{800 \text{ torr } \frac{1 \text{ atm}}{760 \text{ torr}}}$$

$$= 1.2 \times 10^2 \text{ liters}$$

28. 0.759
    [6.10]

The partial pressure of a component gas such as $N_2$ is directly proportional to the mole fraction of that component gas and the total pressure:

$$p_{N_2} = X_{N_2} \, p_{total}$$

The total pressure is the sum of the partial pressures of the components:

$$p_{total} = p_{N_2} + p_{O_2} + p_{CO_2} + p_{H_2O}$$

Therefore,

$$p_{N_2} = p_{total} - p_{O_2} - p_{CO_2} - p_{H_2O}$$

$$= 760 \text{ torr} - 159 \text{ torr} - 0.23 \text{ torr} - 23.8 \text{ torr}$$

$$= 577 \text{ torr}$$

and

$$X_{N_2} = \frac{p_{N_2}}{p_{total}} = \frac{577 \text{ torr}}{760 \text{ torr}} = 0.759$$

29. 4.0 [6.12]

Graham's law of effusion states

$$\frac{r_A}{r_B} = \sqrt{\frac{M_B}{M_A}}$$

Therefore,

$$\frac{3.1}{1} = \sqrt{\frac{38.0}{M_A}}$$

30. $2.6 \times 10^9$ li-
    ters $H_2$ [5.1]

We calculate the number of moles of liquid hydrogen:

$$? \text{ mol } H_2 = 3.4 \times 10^6 \text{ liters } H_2 \left(\frac{10^3 \text{ cm}^3 \text{ } H_2}{1 \text{ liter } H_2}\right)\left(\frac{0.070 \text{ g } H_2}{1 \text{ cm}^3 \text{ } H_2}\right)\left(\frac{1 \text{ mol } H_2}{2.02 \text{ g } H_2}\right)$$

$$= 1.18 \times 10^8 \text{ mol } H_2$$

Then we can calculate the volume of gas at STP:

$$? \text{ liters } H_2 = 1.18 \times 10^8 \text{ mol } H_2 \left(\frac{22.4 \text{ liters } H_2}{1 \text{ mol } H_2}\right)$$

$$= 2.6 \times 10^9 \text{ liters } H_2$$

As the gas at STP, $H_2$ occupies approximately 1000 times the volume of the liquid.

31. $1.0 \times 10^{21}$ li-
    ters $O_2$ [6.7]

At STP each mol of $O_2$, i.e., 32.0 g $O_2$, occupies 22.4 liters. liters. Therefore,

$$? \text{ liters } O_2 = 15 \times 10^{20} \text{ g } O_2 \left(\frac{22.4 \text{ liters } O_2}{32.0 \text{ g } O_2}\right)$$

$$= 1.0 \times 10^{21} \text{ liters } O_2$$

32. 0.996
    [5.2, 6.12]

According to Graham's law,

$$\frac{r_A}{r_B} = \sqrt{\frac{M_B}{M_A}}$$

The molecular weights of the two gases must be calculated:

molecular weight of $^{235}UF_6$ = 235 + 6(19) = 349

molecular weight of $^{238}UF_6$ = 238 + 6(19) = 352

Substituting the molecular weights into the equation of Graham's law, we find

$$\frac{r_{238}UF_6}{r_{235}UF_6} = \frac{349}{352} = \sqrt{0.992} = 0.996$$

33. 4.42 g/liter
    [6.7]

One mole of phosgene weighs 98.9 g, so at STP 98.9 g occupies 22.4 liters. Therefore,

$$\text{density} = \frac{\text{weight}}{\text{volume}}$$

$$= \frac{98.9 \text{ g}}{22.4 \text{ liters}} = 4.42 \text{ g/liter}$$

34. 49.6 g/mol
    [5.1, 6.5]

The ideal gas law can be used. Since $n$, the number of moles, is the weight of material, $g$, divided by the molecular weight, $M$, we can write

$$PV = nRT$$

$$PV = \left(\frac{g}{M}\right)RT$$

$$M = \frac{gRT}{PV}$$

$$= \frac{(0.162 \text{ g})(0.08206 \text{ liter atm K}^{-1} \text{ mol}^{-1})(373 \text{ K})}{(1 \text{ atm})(0.100 \text{ liter})}$$

$$= 49.6 \text{ g/mol}$$

The preceding method is preferred. Any problem in which the temperature, pressure, and volume are not changed can be solved directly with the ideal gas law; problems that involve a change in any of these can be solved easily with a tabular method such as that suggested in problem 1 of this section. The molecular weight of an ideal gas is the mass of gas that would occupy 22.4 liters at STP. Thus, we could solve the problem in the following way:

First, we calculate the volume of 0.162 g of gas at STP:

$$? \text{ ml gas} = 100.0 \text{ ml gas} \left(\frac{273 \text{ K}}{373 \text{ K}}\right)$$

$$= 73.2 \text{ ml gas, or } 0.0732 \text{ liter gas}$$

Then we calculate the density of the gas at STP:

$$\text{density} = \frac{\text{weight}}{\text{volume}} = \left(\frac{0.162 \text{ g}}{0.0732 \text{ liter}}\right) = 2.21 \text{ g/liter}$$

Finally we calculate the molecular weight:

$$M = 2.21 \text{ g/liter} \left(\frac{22.4 \text{ liters}}{1 \text{ mol}}\right) = 49.5 \text{ g/mol}$$

We can also use another method:

$$? \text{ g/mol} = \left(\frac{0.162 \text{ g}}{100 \text{ ml}}\right)\left(\frac{373 \text{ K}}{273 \text{ K}}\right)\left(\frac{1000 \text{ ml}}{1 \text{ liter}}\right)\left(\frac{22.4 \text{ liters}}{1 \text{ mol}}\right)$$

$$= 49.6 \text{ g/mol}$$

Be careful with this method. Volume is in the denominator so the temperature correction factor must be inverted.

35. 1.40 liters
$C_2H_2$
[5.5, 6.9]

First we determine which reagent limits the reaction by calculating the number of moles of each:

$$? \text{ mol CaC}_2 = 4.00 \text{ g CaC}_2 \left(\frac{1 \text{ mol CaC}_2}{64.10 \text{ g CaC}_2}\right)$$

$$= 0.0624 \text{ mol CaC}_2$$

$$? \text{ mol H}_2O = 2.75 \text{ g H}_2O \left(\frac{1 \text{ mol H}_2O}{18.02 \text{ g H}_2O}\right)$$

$$= 0.153 \text{ mol H}_2)$$

Excess water is available since only 0.125 mol $H_2O$ is needed. The limiting reagent is $CaC_2$. Therefore,

$$? \text{ liters C}_2H_2 = 0.0624 \text{ mol CaC}_2 \left(\frac{22.4 \text{ liters C}_2H_2}{1 \text{ mol CaC}_2}\right)$$

$$= 1.40 \text{ liters C}_2H_2$$

36. (a) 5.17 li-
ters $CO_2$
[6.9]

We find

$$? \text{ liters CO}_2 = 3.00 \text{ g C}_2H_2 \left(\frac{(4 \times 22.4) \text{ liters CO}_2}{2(26.02) \text{ g C}_2H_2}\right)$$

$$= 5.17 \text{ liters CO}_2$$

(b) 4.00 li-
ters $CO_2$
[6.9]

You should be able to do part (b) of this problem by inspection. Gay-Lussac's law states that in a chemical reaction the ratio of moles is the same as the ratio of volumes measured at the same temperature and pressure.

37. 1.085 atm
    [6.13]

The van der Waals equation is

$$P + \left(\frac{n^2 a}{V^2}\right)(V - nb) = nRT$$

Rearranging the equation, we find

$$P = \frac{nRT - \left(\frac{n^2 a}{V}\right) + \left(\frac{n^3 ab}{V^2}\right)}{V - nb}$$

Solving for $P$, we find

$$P = 1.085 \text{ atm}$$

From the ideal gas law an answer of 1.092 atm is obtained.

38. $Cl_2$ [6.13]

The value of $a$ is largest for chlorine gas; thus, the intermolecular attraction of chlorine gas molecules is greater than that of the other gas molecules. Notice that the value of $a$ is smallest for helium, the molecules of which show little intermolecular interaction. The value of $a$ is large for molecules with a large dipole and a non-bonding electron pair, such as ammonia molecules. Chlorine gas is also the largest of the gases listed in Table 6.3 of your text; it has the largest value of $b$, a constant which is directly proportional to the molecular radius.

39. 0.23 g $H_2O$
    [6.5, 6.10]

Table 6.2 of your text gives the vapor pressure of water at 25°C:

$$P_{H_2O} = 0.0313 \text{ atm}$$

Even if excess water were added to the vessel, the vapor pressure of water would still be 0.0313 atm. The mass of gaseous water can be calculated from the ideal gas law:

$$PV = nRT = (g/M)RT$$

$$g = \frac{MPV}{RT}$$

$$g = \frac{(18.02 \text{ g/mol})(0.0313 \text{ atm})(10.0 \text{ liter})}{(0.0821 \text{ liter atm } K^{-1} \text{ mol}^{-1})(298°K)}$$

$$= 0.23 \text{ g } H_2O$$

From the density, 1.00 g/ml, we find that only 0.23 ml $H_2O$ evaporated:

$$? \text{ ml } H_2O = 0.23 \text{ g } H_2O \left(\frac{1.00 \text{ ml } H_2O}{1.00 \text{ g } H_2O}\right) = 0.23 \text{ ml } H_2O$$

40. 100 atm [6.10]

The partial pressure of water vapor is 0.0313 atm, and the partial pressure of He is 100 atm. The total pressure is the sum of the two partial pressures:

$$P = p_{H_2O} + p_{He}$$

$$= 0.0313 \text{ atm} + 100 \text{ atm}$$

$$= 100 \text{ atm}$$

41. $2.2 \times 10^3$ K
[5.2, 6.5, 6.14]

First we calculate the number of moles in 1.0 ml of $H_2O(1)$:

$$? \text{ mol } H_2O = 1.0 \text{ ml } H_2O \left(\frac{1.0 \text{ g } H_2O}{1 \text{ ml } H_2O}\right)\left(\frac{1 \text{ mol } H_2O}{18.0 \text{ g } H_2O}\right)$$

$$= 0.0556 \text{ mol } H_2O$$

Then we calculate the temperature from the ideal gas law:

$$T = \frac{PV}{nR} = \frac{(1.0 \text{ atm})(10.0 \text{ liter})}{(0.0556 \text{ mol})(0.0821 \text{ liter atm K}^{-1} \text{ mol}^{-1})}$$

$$= 2.2 \times 10^3 \text{ K}$$

This is above the critical temperature of water, 647.2 K, so no water can exist in liquid form.

42. 25.3 g/mol
[6.5]

Since the gas was weighed before drying, the mass of water vapor must be subtracted to obtain the mass of the unknown gas. We use the ideal gas law to find the mass of the water vapor:

$$? \text{ g } H_2O = \frac{MPV}{RT}$$

$$= \frac{(18.0 \text{ g/mol})(0.0060 \text{ atm})(1.00 \text{ liter})}{(0.0821 \text{ liter atm K}^{-1} \text{ mol}^{-1})(273°\text{K})}$$

$$= 0.0048 \text{ g}$$

The mass of the unknown gas is

$$1.135 \text{ g} - 0.005 \text{ g} = 1.130 \text{ g}$$

The molecular weight of the unknown gas is

$$? \text{ g/mol} = \left(\frac{1.130 \text{ g}}{1.00 \text{ liter}}\right)\left(\frac{22.4 \text{ liters}}{1 \text{ mol}}\right)$$

$$= 25.3 \text{ g/mol}$$

43. $p_{He}$ = 549 torr
and
$p_{N_2}$ = 210 torr
[.9, 6.10]

A mole of the gas mixture would weigh

$$(0.475 \text{ g/liter})(22.4 \text{ liter/mol}) = 10.64 \text{ g/mol}$$

If we let $x$ equal the mole fraction of He and $1 - x$ equal the mole fraction of $N_2$,

$$x(4.00 \text{ g/mol}) + (1 - x)(28.0 \text{ g/mol}) = 10.64 \text{ g/mol}$$

$$x = 0.723$$

$$1 - x = 0.277$$

The partial pressure of helium is the total pressure times the mole fraction of helium:

$$p_{He} = X_{He}P_{total}$$
$$= (0.723)(760 \text{ torr})$$
$$= 549 \text{ torr, or } 0.723 \text{ atm}$$

Similarly for nitrogen:

$$p_{N_2} = X_{N_2}P_{total}$$
$$= (0.277)(760 \text{ torr})$$
$$= 210 \text{ torr, or } 0.277 \text{ atm}$$

**44.** $4.8 \times 10^{21}$ liters gas [6.7]

We find

$$? \text{ liters gas} = 1.0 \times 10^{21} \text{ liters } O_2\left(\frac{1 \text{ liter gas}}{0.21 \text{ liter } O_2}\right)$$
$$= 4.8 \times 10^{21} \text{ liters gas}$$

**45.** larger [6.5]

The volume of the atmosphere would be larger. Atmospheric pressure decreases with height above the earth's surface; thus, the average pressure cannot be 1 atmosphere but a lower value. An average temperature of 273 K is also low, but the effect due to temperature is smaller than that due to pressure.

**46.** 1.09 atm [6.9]

Since this problem contains so many variables, it is best to tabulate the information:

|   | Initial Conditions | Final Conditions |
|---|---|---|
| $P$ | 1.01 atm | ? |
| $V$ | 1.47 liters | 1.47 liters |
| $T$ | (25 + 273) K | (54 + 273) K |
| $n$ | ? $n_{O_2}$ | ? $(n_{O_2} + n_{O_3})$ |

We use the ideal gas law, $PV = nRT$, to determine the initial number of moles of $O_2$:

$$n = \frac{PV}{RT} = 6.07 \times 10^{-2} \text{ mol } O_2$$

We then calculate the number of moles of $O_2$ that react (5.24% of the original amount);

$$n_{O2} \text{ reacting} = (0.0524)(6.07 \times 10^{-2} \text{ mol } O_2)$$
$$= 3.18 \times 10^{-3} \text{ mol } O_2$$

and the number of moles of $O_2$ remaining:

$$n_{O_2} \text{ remaining} = (6.07 \times 10^{-2} \text{ mol}) - (3.18 \times 10^{-3} \text{ mol})$$

$$= 5.75 \times 10^{-2} \text{ mol } O_2$$

We can use the number of moles of $O_2$ reacting and the balanced equation to determine the number of moles of $O_3$ formed:

$$n_{O_3} = 3.18 \times 10^{-3} \text{ mol } O_2 \left( \frac{2 \text{ mol } O_3}{3 \text{ mol } O_2} \right)$$

$$= 2.12 \times 10^{-3} \text{ mol } O_3$$

The total amount of gas remaining after the reaction is

$$n_{total} = n_{O_2} + n_{O_3}$$

$$= (5.75 \times 10^{-2} \text{ mol } O_2) + (2.12 \times 10^{-3} \text{ mol } O_3)$$

$$5.96 \times 10^{-2} \text{ mol gas}$$

From $PV = nRT$ we find the final pressure:

$$PV = nRT$$

$$= \frac{(5.96 \times 10^{-2} \text{ mol})(0.08206 \text{ liter atm mol}^{-1} \text{ K}^{-1})(327°K)}{1.47 \text{ liters}}$$

$$= 1.09 \text{ atm}$$

## SELF-TEST

Complete the test in 50 minutes:

I. Work the following problems:

1. A bulb filled with a gas to a pressure of 760 mm weighs 116.3124 g. When the bulb is heated to 88°C at a pressure of 760 mm, 100 ml of gas is expelled and the bulb and remaining gas weight 116.2584 g. What is the molecular weight of the gas?

2. A 1.168-g sample of an oxide, $XO_2$, reacts with exactly 500 ml of $H_2$ gas measured at STP. What is the molecular weight of $XO_2$?

$$XO_2(s) + 2H_2(g) \rightarrow X(s) + 2H_2O(g)$$

3. A 100-ml sample of gas is collected over water at temperature $T_1$, and the wet gas is found to exert a pressure of 750 mm at that temperature. The same sample of gas is found to occupy 97.1 ml at $T_1$ when dry and under a pressure of 760 mm. Calculate the vapor pressure of water at $T_1$ from these data.

4. In a mixture of $C_2H_6$ and $O_2$ confined in a 1.00 liter container, the partial pressure of $C_2H_6$ is 160 mm and the partial pressure of $O_2$ is 560 mm. The

mixture is ignited and reacts according to the equation:

$$2C_2H_6(g) + 7O_2(g) \rightarrow 4CO_2(g) + 6H_2O(g)$$

The temperature is constant throughout the experiment and is high enough so that the $H_2O$ is gaseous. What is the total pressure of the final mixture?

5. Calculate the density of a gas in g/liter at STP if a given volume of the gas effuses through an apparatus in 5.00 minutes and the same volume of oxygen at the same temperature and pressure effuses through the apparatus in 6.30 minutes.

6. Assume that 10 liters of $NH_3(g)$ and 10 liters of $O_2(g)$ are mixed and react according to the equation:

$$4NH_3(g) + 5O_2(g) \rightarrow 4NO(g) + 6H_2O(g)$$

If the conditions under which the gases are measured are constant and such that all materials are gaseous, list the volumes of all materials at the conclusion of the reaction.

7. Calculate the molecular weight of a gas that has a density of 1.59 g/liter at 50°C and a pressure of 730 mm.

II. Complete the statement or answer the question:

1. Of the two gases, $H_2$ (molecular weight = 2) and $C_5H_{12}$ (molecular weight = 72), under the same conditions of temperature and pressure, _____ would effuse _____ times more rapidly through a given orifice.

2. If 1.00-liter samples of $H_2(g)$ and $C_5H_{12}(g)$ are considered, both at STP, which sample has the larger number of molecules? _____ The molecules of which sample have the larger average kinetic energy?

3. The partial pressure of oxygen in a flask containing 32 g of $O_2$ and 32 g of $H_2$ is _____ of the total pressure. (Atomic weights: O = 16.0; H = 1.0.)

4. The behavior of real gases deviates from that described by the ideal gas law because the ideal gas law fails to take into account _____ and _____ .

III. Using the coordinates given, sketch the approximate shape of the curve for an ideal gas:

a. pressure vs. volume

c. absolute temperature vs. volume

b. pressure vs. product of pressure and volume

d. energy distribution of molecules

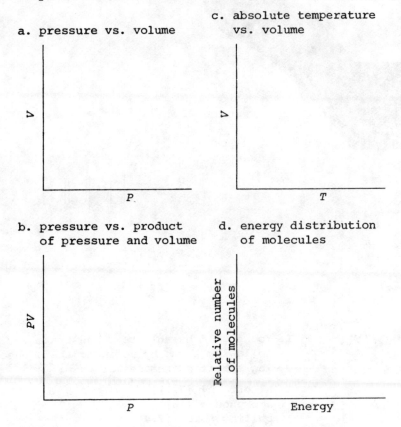

# Liquids and Solids

OBJECTIVES    I. You should be able to demonstrate your knowledge of the
following terms by defining them, describing them, or giving specific examples of them:

allotrophy [7.11]
amorphous [7.6]
boiling point [7.4]
Bragg equation [7.10]
Clausius-Clapeyron equation [7.5]
crystal lattice structures [7.9, 7.11, 7.12]
dislocations [7.14]
entropy of vaporization [7.5]
lattice energy [7.13]
molar entropy of fusion [7.6]
molar heat of sublimation [7.8]
molar heat of vaporization [7.2, 7.5]
$n$-type [7.14]
normal boiling point [7.4]

normal freezing point [7.6]
*p*-type [7.14]
point defects [7.14]
semiconductor [7.14]
sublimation [7.8]
supercooling [7.6]
surface tension [7.1]
triple point [7.8]
Trouton's rule [7.5]
unit cell [7.9]
vapor pressure [7.3, 7.7]
viscosity [7.1]

II. You should be able to determine the heat of vaporization, $\Delta H_V$, from the vapor pressure of a liquid at two temperatures and rearrange the Clausius-Clapeyron equation to determine either vapor pressure or temperature if given the other variables.

III. You should be able to interpret phase diagrams.

IV. You should be familiar with the various types of crystal lattices and be able to calculate densities from structural data and vice versa.

V. You should be able to use the Bragg equation to determine the distance between diffraction planes.

## EXERCISES

I. Answer each of the following with an entry from the list on the right:

1. Water can exist at 100°C and 1 atm as either gas or ____.
2. The phase in which molecular motion is most restricted in the ____ phase.
3. As a liquid is heated, the vapor pressure ____.
4. ____ is a property of liquids that describes resistance to flow.
5. In the ____ state matter assumes the shape of and completely fills the container into which it is placed.

a. decreases(s)
b. increase(s)
c. remains the same
d. gas
e. liquid
f. solid
g. surface tension

6. A drop of liquid has a spherical shape due to the property of ____.

7. As pressure is decreased, boiling points ____.

8. As the temperature of a liquid increases, the rate of evaporation ____.

9. ____ is the condition in which the rates of two opposite tendencies are equal.

10. The pressure of vapor in equilibrium with a liquid at a given temperature is called ____.

11. At the ____ the vapor pressure of a liquid equals 1 atmosphere of pressure.

12. The equation

$$\frac{\Delta H_v}{T_b} = +88 \text{ J/K mol}$$

is a statement of ____.

13. The ____ is the temperature at which liquid and solid are in equilibrium under 1 atmosphere of pressure.

14. ____ is the process in which a solid passes directly to the vapor phase.

15. An ____ semiconductor material contains an excess number of electrons.

16. The ____ cubic crystal is the cubic crystal that has the most empty space.

17. Each atom has six nearest neighbors in the ____ cubic crystal.

18. There are two atoms per unit cell in the ____ cubic crystal.

19. The ____ cubic crystal is the most densely packed cubic crystal.

h. viscosity

i. Hess's law

j. Trouton's rule

k. melting point

l. normal boiling point

m. sublimation

n. equilibrium

o. vapor pressure

p. *n*-type

q. *p*-type

r. simple

s. body-centered

t. face-centered

20. The diagram

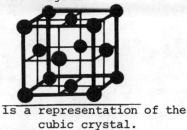

is a representation of the
_____ cubic crystal.

II. Sulfur can exist in several crystalline forms. The phase
diagram in Figure 7.1(a) is that for monoclinic sulfur.

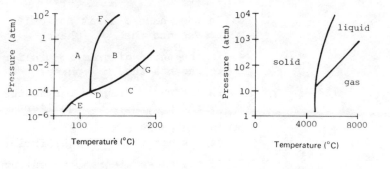

(a)                              (b)

FIGURE 7.1    Phase diagrams for (a) monoclinic sulfur
and carbon.

The unit cell of monoclinic sulfur does not have any equal
edge dimensions and one of the angles is not 90°:

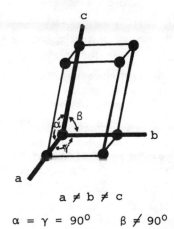

$a \neq b \neq c$

$\alpha = \gamma = 90°$        $\beta \neq 90°$

Solid carbon can exist as diamond and as graphite. Diamond can only be formed at pressures above $10^4$ atmospheres; after it is formed, it cannot easily be destroyed at atmospheric pressures even though graphite is the preferred structure.

Use the information in the phase diagrams of Figure 7.1 and answer each of the following questions with an entry from the list on the right:

1. In the phase diagram for sulfur area A represents the _____ phase.

2. In the phase diagram for sulfur area B represents the _____ phase.

3. In the phase diagram for sulfur area C represents the _____ phase.

4. In the phase diagram for sulfur area D represents the _____ phase.

5. In the phase diagram for sulfur line E represents the _____ equilibrium.

6. In the phase diagram for sulfur line F represents the _____ equilibrium.

7. In the phase diagram for sulfur line G represents the _____ equilibrium.

8. Which element has the lower melting point: sulfur or carbon? _____

9. When sulfur is heated from 100°C to 150°C under 1 atmosphere of pressure, a _____ phase change occurs.

10. When carbon is heated from 100°C to 150°C under 1 atmosphere of pressure, a _____ phase change occurs.

11. As the pressure exerted on a sample of sulfur maintained at 100°C is reduced from 1 atmosphere to $10^{-6}$ atmosphere, a _____ phase change occurs.

12. As the pressure exerted on a sample of of sulfur maintained at 150°C is reduced from 1 atmosphere to $10^{-6}$ atmosphere, a _____ phase change occurs.

13. Which carbon phase exists at 3000°C and 1 atmosphere pressure? _____

a. sulfur

b. carbon

c. sulfur and carbon

d. gas-liquid

e. gas-solid

f. liquid-solid

g. liquid-gas

h. solid-gas

i. solid-liquid

j. none

k. gas

l. liquid

m. solid

n. triple point

14. Which element never exists as a liquid at 1 atmosphere of pressure: carbon or sulfur?

15. Which is more dense: liquid or solid sulfur? ____

III. Identify the phase(s) present in each region of the following curve and discuss the energy changes occurring as a solid is heated from point 1 to point 6:

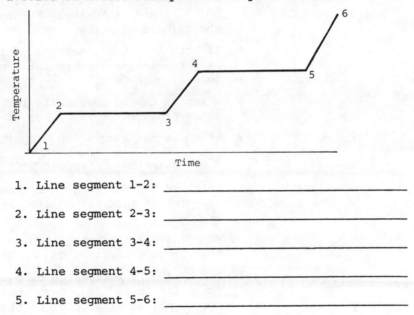

1. Line segment 1-2: _____

2. Line segment 2-3: _____

3. Line segment 3-4: _____

4. Line segment 4-5: _____

5. Line segment 5-6: _____

IV. Answer each of the following:

1. Argon crystallizes in a face-centered cubic structure at −189°C. If the density of the solid is 1.7 g/cm$^3$, calculate the edge length of a unit cell.

2. An oxide of zirconium forms a face-centered cubic lattice of zirconium ions, and the tetrahedral holes are occupied by oxide ions. What is the formula of the zirconium oxide?

3. Crystal studies have shown that a certain sulfide of manganese forms a face-centered cubic lattice with sulfide ions filling the octahedral holes. What is the formula of the compound?

4. In the mineral spinel one-eighth of the tetrahedral holes are occupied by $Mg^{2+}$ ions, and one-half of the octahedral holes are occupied by $Al^{3+}$ ions. Oxide ions occupy the corners and face-centers of the unit cell. What is the formula for spinel?

5. The mineral perovskite has a crystalline structure in which the oxide ions occupy the face-centers and the larger cations, $Ca^{2+}$, occupy the corners of the unit cell. The smaller cations, $Ti^{4+}$, occupy the octahedral holes formed exclusively by the oxide ions. What is the formula of perovskite?

6. If both form the same type of lattice, which would have the larger electrostatic potential energy per mole: MgO or $CuSO_4$?

7. Which of the following liquids should exert the smaller vapor pressure at 25°C: $CCl_4$ or $H_2O$?

8. Which is the largest for a particular substance:

$\Delta H_{vaporization}$, $\Delta H_{fusion}$, or $\Delta H_{sublimation}$

9. Which of the following ionic compounds should have the highest melting temperature: $CaSO_4$, CaS, MgO, or $NH_4NO_3$?

10. Which of the following liquids should have the smaller heat of vaporization: $NH_3$ or $BCl_3$?

11. The heat of vaporization of $S_2Cl_2$ at the normal boiling point is 267 J/g. Use Trouton's rule to predict hhe normal boiling point of this compound.

12. In the diffraction of a gold crystal with X rays of a wavelength equal to 1.54 Å, a first-order reflection is shown at an angle of 22°10 . What is the distance between the diffracted planes?

13. Aluminum crystallizes in a face-centered cubic unit cell with the length of an edge equal to 4.05 Å. Assume the atoms are hard spheres and each face-centered atom touches the four corner atoms of its face. Calculate the radius of a hard-sphere atom.

ANSWERS TO
EXERCISES

I. Physical states of matter

1. liquid [7.4]
2. solid [7.7]
3. increases [7.3]

4. viscosity [7.1]
5. gas [6.1]
6. surface ten-
   sion [7.1]
7. decreases [7.4]
8. increases [7.2]
9. equilibrium
   [7.3]
10. vapor pressure
    [7.3]
11. normal boiling
    point [7.4]
12. Trouton's rule
    [7.5]
13. melting point
    [7.6]
14. sublimation
    [7.8]
15. *n*-type [7.14]
16. simple [7.9]
17. simple [7.9]
18. body-centered
    [7.9]
19. face-centered
    [7.11]
20. face-centered
    [7.9]

It is important to remember the terms that apply to phase changes. The following diagram may be of help:

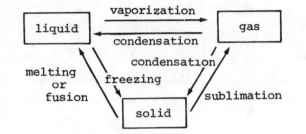

II. Phase diagrams for monoclinic sulfur and carbon.

See Section 7.8 of your text for a discussion of phase diagrams.

1. solid
2. liquid
3. gas
4. triple point
5. gas-solid or
   solid-gas
6. liquid-solid
   or solid-solid
7. gas-liquid or
   liquid-gas

8. sulfur

Carbon does not form a liquid phase at 1 atmosphere; it sublimes when heated to 3652°C.

9. solid-liquid

10. none

11. solid-gas

12. liquid-gas

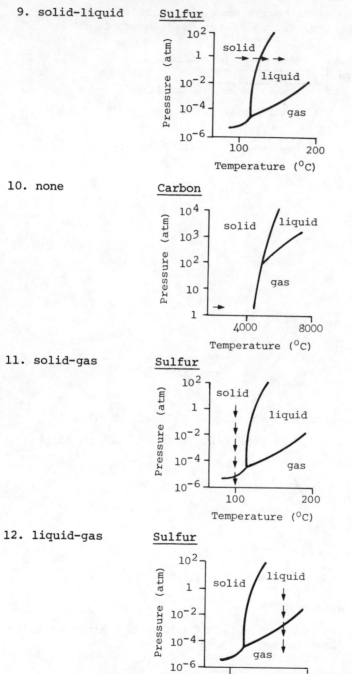

13. solid

### Carbon

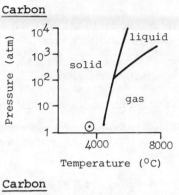

14. carbon

### Carbon

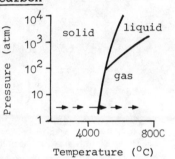

15. solid

As pressure is increased, the material becomes more dense, and the solid form is produced:

### Sulfur

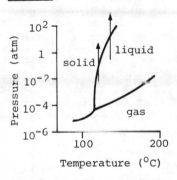

III. Phase and energy changes

See Sections 7.4 and 7.6 of your text.

1. Line segment
   1-2:
   In this region the solid phase is present. The temperature and kinetic energy of the molecules are increasing.

2. Line segment
   2-3:
   Solid and liquid are in equilibrium at the melting point. The temperature and the kinetic energy are constant. The potential energy increases, because the system is still being heated.

3. Line segment
   3-4:
   In this region the liquid is present. The temperature and the kinetic energy are increasing.

4. Line segment   Liquid and vapor are in equilibrium at the boiling point.
   4-5:             The temperature and the kinetic energy are constant. The
                    potential energy is increasing.

5. Line segment   In this region the gas phase is present. The temperature
   4-5:             and kinetic energy are increasing.

### IV. Molecular interactions in solids and liquids

1. $5.4 \times 10^{-8}$ cm,   Argon forms a face-centered cubic crystal, and there are
   or 5.4 Å                    four atoms in a unit cell. First we calculate the mass of
   [7.9]                       the four atoms:

$$? \text{ g Ar} = 4 \text{ Ar atoms} \left(\frac{1 \text{ mol Ar}}{6.022 \times 10^{23} \text{ atoms Ar}}\right)\left(\frac{39.948 \text{ g Ar}}{1 \text{ mol Ar}}\right)$$

$$= 2.653 \times 10^{-22} \text{ g Ar}$$

Then we use the density to calculate the volume occupied
by the unit cell:

$$? \text{ cm}^3 = 2.653 \times 10^{-22} \text{ g Ar}\left(\frac{1 \text{ cm}^3}{1.7 \text{ g Ar}}\right)$$

$$= 1.56 \times 10^{-22} \text{ cm}^3$$

Finally, we calculate the edge of the unit cell:

$$? \text{ cm} = \sqrt[3]{\text{cm}^3} = \sqrt[3]{1.56 \times 10^{-22} \text{ cm}^3} = \sqrt[3]{160 \times 10^{-24} \text{ cm}^3}$$

$$= 5.4 \times 10^{-8} \text{ cm}$$

2. $ZrO_2$       The zirconium ions are in a face-centered cubic lattice,
   [7.12]         and there are four zirconium ions per unit cell. There
                  are eight tetrahedral holes; thus, there are eight oxide
                  ions. The simplest formula is $ZrO_2$.

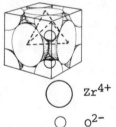

$Zr^{4+}$

$O^{2-}$

A tetrahedral hole is located directly behind a corner
in a face-centered cubic unit cell. The tetrahedron is
formed with a corner atom and the atoms in the centers
of the faces that form the corner. A tetrahedral hole is
shown in the diagram to the left. There are eight tetrahedral holes in each cell.

3. MnS          The manganese ions form a face-centered cubic lattice,
   [7.12]        and there are four manganese ions per unit cell. There
                  are four octahedral holes; thus, there are four sulfide
                  ions. The simplest formula is MnS.

One-fourth of an octahedral hole is shown in the following
diagram of the unit cell. There are twelve such quarter
holes in a unit cell. One complete octahedral hole is in
the center of the unit cell.

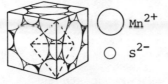

4. MgAl$_2$O$_3$
[7.12]

The Mg$^{2+}$ ions occupy one-eighth of the tetrahedral holes. Since there are eight tetrahedral holes, there is (1/8)(8), or one Mg$^{2+}$ ion.

There is one octahedral hole in the center and one on each edge. Each edge is shared by four unit cells, so there are (1/4)(12), or three edge octahedral holes per unit cell. The total number of octahedral holes per unit cell is therefore four. The Al$^{3+}$ ions occupy one-half of the octahedral holes; thus, there are (1/2)(4), or two Al$^{3+}$ ions in a cell.

Oxide ions occupy the corners and face centers. There are eight corners, each shared by eight unit cells; thus, there is (1/8)(8), or one oxide ion per unit cell. There are six faces, each shared by two unit cells; thus, there are (1/2)(6), or three oxide ions per unit cell.

The formula of spinel is therefore MgAl$_2$O$_3$.

5. CaTiO$_3$

There are eight corner Ca$^{2+}$ ions, each shared by eight unit cells; thus, there is (1/8)(8), or one Ca$^{2+}$ ion per unit cell.

There are six face-center positions containing an oxide ion, each shared by another unit cell; thus, there are (1/2)(6), or three oxide ions per unit cell.

The Ti$^{4+}$ ion is in the center of the unit cell, in the one octahedral hole formed exclusively by oxide ions. It is not shared with any other unit cell, so there is one Ti$^{4+}$ ion per unit cell.

The formula of perovskite is therefore CaTiO$_3$.

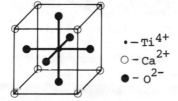

6. MgO

The equation for calculating potential energy is

$$PE = -\left(\frac{(Z_c)(Z_a)e^2}{r}\right)NA$$

For both MgO and $CuSO_4$,

$$Z_a = 2$$

$$Z_c = 2$$

and $e^2$ and $N$ are constants.

The Madelung constant, $A$, is the same for both compounds since they have the same crystalline structure. Since $r$ is the only variable and since it is in the denominator, the numerical value of $PE$ would be greater for whichever compound has the smaller value of $r$, the ionic radius. From the positions in the periodic table, we would expect $Mg^{2+}$ to be slightly smaller than $Cu^{2+}$. Also, we would expect $O^{2-}$ to be much smaller than $SO_4^{2-}$. Thus, since

$$r_{MgO} < r_{CuSO_4}$$

we would expect

$$|PE_{MgO}| > |PE_{CuSO_4}|$$

7. $H_2O$
   [7.3]

Liquids with strong intermolecular attractive forces have low vapor pressures. Carbon tetrachloride is a nonpolar, tetrahedral molecule and water is a polar, bent molecule:

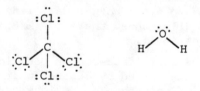

The only attractive forces that exist between $CCl_4$ molecules are relatively weak van der Waals forces; however, there are rather strong dipole-dipole attractive forces between $H_2O$ molecules. Thus, the vapor pressure of water is lower than that of carbon tetrachloride.

8. $\Delta H_{sublimation}$
   [5.9, 7.5, 7.6]

Sublimation is the process in which a solid goes directly to the vapor state; it requires more energy than a phase change from solid to liquid or liquid to vapor. At any given temperature

$$\Delta H_{sublimation} = \Delta H_{fusion} + \Delta H_{vaporization}$$

We obtain the values of the heats of fusion and vaporization for water at 0°C from thermodynamic tables and use them with the law of Hess to calculate the heat of sublimation of water at 0°C (see Section 5.8 of your text):

$$H_2O(s) \xrightarrow{0°C} H_2O(l) \qquad \Delta H_f = 5.98 \text{ J/mol}$$
$$H_2O(l) \xrightarrow{0°C} H_2O(g) \qquad \Delta H_v = 198.74 \text{ J/mol}$$

$$H_2O(s) \xrightarrow{0°C} H_2O(g) \quad \Delta H_{subl.} = 204.72 \text{ J/mol}$$

9. MgO

The substance with the highest melting point is that with the strongest intermolecular attractive forces; i.e., higher temperatures are required to break the stronger lattice. The greater the value of the electrostatic potential energy, the more difficult it is to break the lattice:

$$PE = - \frac{(Z_c)(Z_a)e^2}{r} NA$$

In the equation $e^2$ and $N$ are constants, and the values of $A$ for the series of compounds given in this problem are similar (see Table 7.4 of your text). The compounds have cesium chloride, $A = 1.763$, sodium chloride, $A = 1.748$, or zinc blende, $A = 1.638$, structures. None could have fluorite or rutile. Of the compounds in the series, $NH_4NO_3$ is the only one with ionic charges of 1+ and 1-; this fact eliminates $NH_4NO_3$ as a possibility, because all others have $Z_c = Z_a = 2$. To decide among the remaining three substances, we consider $r$. Since $r$ is in the denominator, the smallest value of $r$ will give the largest potential energy, $PE$. The $Mg^{2+}$ ion is smaller than $Ca^{2+}$, and $O^{2-}$ is smaller than either $S^{2-}$ or $SO_4^-$; thus, MgO has the smallest value of $r$ and the largest value of $PE$.

10. $BCl_3$

In general, the lower the heat of vaporization, the weaker the intermolecular forces of attraction. Boron trichloride, $BCl_3$, is a nonpolar molecule with attractive forces weaker than those of the polar ammonia molecule, $NH_3$.

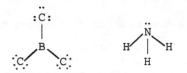

11. 140°C
    [7.5]

Trouton's rule states

$$\frac{\Delta H_v}{T_b} = 88 \text{ J/mol}$$

The molar heat of vaporization is calculated:

$$\Delta H_v = 267 \text{ J/g} \, \frac{135.0 \text{ g } S_2Cl_2}{1 \text{ mol } S_2Cl_2} = 361 \times 10^4 \text{ J/mol}$$

Therefore,

$$T_b = \frac{3.61 \times 10^4 \text{ J/mol}}{88 \text{ J/K mol}} = 410 \text{ K}$$

Changing $T_b$ from Kelvin to centigrade, we find

$$T_b = 140°C$$

12. 2.04 Å
[7.10]

We substitute values into the Bragg equation:

$$n\lambda = 2d \sin \phi$$

$$1(1.54 \text{ Å}) = 2(d)(0.377)$$

To determine sin 22°10', we look up the value in a table or use a slide rule or calculator. The value, 0.377, was obtained on a slide rule.

We then solve the Bragg equation:

$$d = \frac{1.54 \text{ Å}}{2(0.377)} = 2.04 \text{ Å}$$

13. 1.43 Å
[7.9]

This is a problem of geometry. Envision the unit cell:

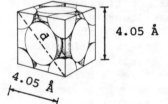

4.05 Å

4.05 Å

Using the Pythagorean theorem, we calculate the length of the hypotenuse, $d$:

$$d = \sqrt{2}(4.05 \text{ Å}) = 5.73 \text{ Å}$$

The length $d$ corresponds to 4 radii; thus, $r = d/4$ = 1.43 Å. The value of $r$ is slightly larger than expected (see Figure 3.1 of your text).

SELF-TEST

I. Complete the test in 20 minutes:

1. Complete the phase diagram of $H_2O$. Label the triple point, the normal boiling point, the normal freezing point, and the solid, liquid, and vapor phases. Clearly show the approximate slopes of all lines.

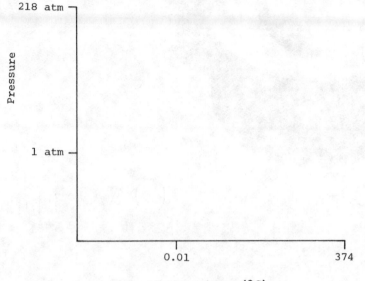

Temperature ($^o$C)

2. How many atoms are in the unit cell of
   (a) a simple cubic lattice
   (b) a body-centered cubic lattice

3. At 328°K the vapor pressure of a liquid is 0.823 atm.
   If the heat of vaporization is 7.07 kcal/mol, what is
   the normal boiling point of the liquid? Do not use
   Trouton's rule.

4. What is the atomic weight of an element that crystallizes as a face-centered cubic crystal, the density of
   which is 8.94 g/cm$^3$? The length of the diagonal
   through the center of the cube if 6.10 Å.

# Oxygen and Hydrogen

OBJECTIVES

I. You should be able to demonstrate your knowledge of the following terms by defining them, describing them, or giving specific examples of them:

acidic oxides [8.3]
allotrophy [8.4]
amphoterism [8.3]
basic oxides [8.3]
deuterium [8.7]
displacement [8.6]
disproportionation [8.5]
electrolysis [8.6]
hydride [8.7]
hydrogen bond [8.8]
hydronium ion [8.7]
hydroxide [8.3]
interstitial [8.7]

ion-electron method [8.5]
oxidation [8.5]
oxide ion [8.1, 8.2]
oxidizing agent [8.5]
oxidation—reduction method [8.5]
ozone [8.4]
peroxide ion [8.1, 8.2]
reducing agent [8.5]
reduction [8.5]
superoxide [8.2]
tritium [8.7]
water gas [8.6]

II. You should be familiar with the chemical and physical properties of oxygen and hydrogen.

III. You should be familiar with simple reactions of oxygen and/or hydrogen with elements and compounds. Your instructor will emphasize the important reactions.

IV. You should be able to balance oxidation-reduction equations by using the oxidation-number method and/or the ion-electron method.

EXERCISES

I. In this section you will find questions that involve concepts and methods presented in Chapters 1 through 8 of your text. These questions are designed to help you meet the objectives of Chapter 0 and to serve as a brief review. Answer each of the following:

1. Is the following statement *true* or *false*? The density of hydrogen gas is lower than that of any other gas.

2. Hydrogen gas can be prepared commercially by each of the following methods. Balance the equations:

(a) $C(s) + H_2O(g) \xrightarrow{1000°C} CO(g) + H_2(g)$

(b) $H_2O(l) \xrightarrow{electrolysis} H_2(g) + O_2(g)$

(c) $Fe(s) + H_2O(g) \xrightarrow{400°C} Fe_3O_4(s) + H_2(g)$

(d) $Zn(s) + H^+(aq) \rightarrow Zn^{2+}(aq) + H_2(g)$

3. Determine the oxidation number of each atom in problem 2 of this section.

4. Identify the reducing agent in each reaction in problem 2 of this section.

5. Which compound would you expect to have a higher normal boiling point: $O_2$ or $O_3$?

6. Draw a Lewis structure of the peroxide ion, $O_2^{2-}$, and the superoxide ion, $O_2^-$.

7. Write the molecular-orbital configuration of $O_2^{2-}$.

8. What is the bond order of the superoxide anion, $O_2^-$.

9. Which type of magnetic behavior would you expect for a solid containing the superoxide anion?

10. How many protons are in the nucleus of an oxygen atom?

11. Write the quantum numbers of each electron of an oxygen atom.

12. Which type of crystalline solid would oxygen form?

13. Which of the following molecules should contain the least polar bond: $OF_2$, $H_2O_2$, or $SO_2$?

14. Yellow phosphorus is very toxic and is used in some rat poisons. Since elemental phosphorus is normally not a constituent of biological material, the following nonquantitative screening test can be used to detect it:

Elemental phosphorus, $P_4$, is converted in the presence of water to hypophosphorous acid, $H_3PO_2$, and phosphine, $PH_3$ (reaction 1). The hypophosphorous acid converts silver ions to metallic silver (reaction 2). The phosphine reacts with the silver ions of $AgNO_3$ to form silver phosphide, $Ag_3P$ (reaction 3). Phosphine and silver phosphide each give a brown stain to filter paper impregnated with $AgNO_3$.

Do the following:

(a) Write a balanced equation for each reaction of the test.
(b) Identify theoxidizing and reducing agents in each redox reaction of the test.

15. Balance each of the following equations. All reactions occur in acid solution.

(a) $MnO_4^- + H_2C_2O_4 \rightarrow Mn^{2+} + CO_2$

(b) $Sn^{2+} + HgCl_2 \rightarrow Hg_2Cl_2 + Sn^{4+} + Cl^-$

(c) $MnO_4^- + Mo^{3+} \rightarrow MoO_4^{2-} + Mn^{2+}$

(d) $Ce^{4+} + H_3AsO_3 \rightarrow Ce^{3+} + H_3AsO_4$

16. Balance each of the following equations. All reactions occur in alkaline solution.

(a) $IO_4^- + I^- \rightarrow IO_3^- + I_2$

(b) $CO(NH_2)_2 + OBr^- \rightarrow CO_2 + N_2 + Br^-$

(c) $Al \rightarrow AlOH_4^- + H_2$

## II. Oxygen

On earth, oxygen composes an estimated 46.5 percent of the crust, 89 percent of the water, 23.2 percent of the air, and possibly as much as 50 percent of the minerals (all percentages are by weight). Living organisms on earth contain approximately one oxygen atom for every three other atoms. The reactions of oxygen are numerous and of great importance. We will consider only a few examples of such reactions. Answer the following:

1. Oxygen in the atmosphere reacts in the presence of ultraviolet light to produce ozone, $O_3$. The ozone absorbs most of the ultraviolet light in the upper atmosphere and protects the earth from the damaging rays. Write a balanced equation for the production of ozone and give the oxidation number of oxygen in ozone.

2. Oxygen is consumed in plant and animal respiration, and carbon dioxide and water are produced. For example,

$$C_6H_{12}O_6(aq) + O_2(g) \rightarrow CO_2(g) + H_2O(l)$$
glucose

$$C_{12}H_{22}O_{11}(aq) + 12O_2(g) \rightarrow 12CO_2(g) + 11H_2O(l)$$
maltose

The $CO_2$ that is produced enters the carbon dioxide cycle, and the $H_2O$ enters the water cycle. Carbon dioxide dissolves in the oceans where it can react with $Ca^{2+}$ to form carbonate deposits:

$$Ca^{2+}(aq) + H_2O(l) + CO_2(g) \rightarrow CaCO_3(s) + 2H^+(aq)$$

Photosynthesis by the phytoplankton in the sea returns oxygen to the atmosphere:

$$6CO_2(g) + 6H_2O(l) \rightarrow 6O_2(g) + C_6H_{12}O_6(aq)$$

Which of the preceding processes would irreversibly reduce the amount of $O_2$ in the atmosphere under ordinary conditions?

3. Rocks exposed to the atmosphere can be oxidized by oxygen; for example, iron(II) oxide is oxidized to iron(III) oxide. Write the equation for the atmospheric oxidation of iron(II) oxide.

4. Oxygen oxidizes many compounds and elements. The fuels we burn consume oxygen and produce $CO_2$. Write a balanced equation for the complete oxidation of coal (assume it is only carbon). Also write a balanced equation for the complete oxidation of $C_8H_{18}$, octane.

5. Oxygen oxidizes the volcanic gases CO, $SO_2$, and $H_2$. Write a balanced equation for the complete oxidation of each of the volcanic gases.

III. Hydrogen

Hydrogen composes 63 percent of the total number of atoms in living organisms on the earth and 91 percent of the number of atoms in our universe. The ability of certain molecules to form hydrogen bonds is important. Without hydrogen bonding life would be different; for example, water would boil at a much lower temperature, and DNA would not be a double helix.

1. The mass of the earth's atmosphere is approximately $5 \times 10^5$ tons (1 ton = 2000 lb). If 0.5 percent of the mass is $H_2$, how many tons of $H_2$ are in the atmosphere?

2. The earth contains approximately $1.5 \times 10^{24}$ g of free water. How many tons of hydrogen combined with oxygen to form the earth's free water?

ANSWERS TO EXERCISES

I. Review

1. True
   [6.7, 8.7]

If we consider hydrogen to be an ideal gas, 1 mol of that gas would occupy 22.4 liters. Since 1 mol of $H_2$ is 2 g of $H_2$, there would be 2 g of $H_2$ in 22.4 liters. Similarly, the density of helium would be 4 g/22.4 liters. All other gases are heavier than hydrogen or helium and would have a larger density.

2., 3., 4.
   [3.12, 8.5]

Oxidation numbers are given above the balanced equations, and reducing agents are identified by the letter $R$ below the equation:

$$\overset{0}{C}(s) + \overset{1+\ 2-}{H_2O}(g) \xrightarrow{1000°C} \overset{2+\ 2-}{CO}(g) + \overset{0}{H_2}(g)$$

(a)   $R$

(b) $\underset{R}{2H_2O(l)}$ $\xrightarrow{\text{electrolysis}}$ $2H_2(g) + O_2(g)$

(with oxidation states: $1+$ $2-$ over $H_2O$; $0$ over $H_2$; $0$ over $O_2$)

(c) $\underset{R}{3Fe(s)} + 4H_2O$ $\xrightarrow{400°C}$ $Fe_3O_4(s) + 4H_2(g)$

(with oxidation states: $1+$ $2-$ over $H_2O$; $8/3+$ $2-$ over $Fe_3O_4$; $0$ over $H_2$)

(d) $\underset{R}{Zn(s)} + 2H^+(aq) \rightarrow Zn^{2+}(aq) + H_2(g)$

(with oxidation states: $0$ over $Zn$; $1+$ over $H^+$; $2+$ over $Zn^{2+}$; $0$ over $H_2$)

5. $O_3$
   [4.7, 8.4]

   The molecular weight of $O_3$ is greater than that of $O_2$, and the attractive forces of $O_3$, which has a dipole moment, are stronger than those of $O_2$, which has no dipole moment.

6. $[:\overset{..}{\underset{..}{O}}:\overset{..}{\underset{..}{O}}:]^{2-}$

   $[:\overset{..}{\underset{..}{O}}:\overset{..}{O}\cdot]^{-}$

   [3.7]

7. [4.5]

   The molecular-orbital configuration of $O_2^{2-}$ is

   $(\sigma 1s)^2(\sigma*1s)^2(\sigma 2s)^2(\sigma*2s)^2(\sigma 2p)^2(\pi 2p)^4(\pi*2p)^4$.

8. $1\frac{1}{2}$
   [4.5, 8.2]

9. paramagnetic
   [4.5]

   There is one unpaired electron in $O_2^-$.

10. 8 protons
    [2.6, 2.7]

11. [2.13]

| Electron | $n$ | $l$ | $m$ | $s$ |
|----------|-----|-----|-----|-----|
| 1 | 1 | 0 | 0 | $+\frac{1}{2}$ |
| 2 | 1 | 0 | 0 | $-\frac{1}{2}$ |
| 3 | 2 | 0 | 0 | $+\frac{1}{2}$ |
| 4 | 2 | 0 | 0 | $-\frac{1}{2}$ |
| 5 | 2 | 1 | +1 | $+\frac{1}{2}$ |
| 6 | 2 | 1 | 0 | $+\frac{1}{2}$ |
| 7 | 2 | 1 | -1 | $+\frac{1}{2}$ |
| 8 | 2 | 1 | +1 | $-\frac{1}{2}$ |

12. [4.9]

    Oxygen would form a nonpolar molecular solid with van der Waals forces holding the molecules together.

13. $H_2O_2$
    [3.10]

    In peroxide the O—O is nonpolar.

14. [8.5]
    (a)             Reaction (1),

    $P_4 + 6H_2O \rightarrow 3H_3PO_2 + PH_3$

                is a disproportionation reaction:

$$3[P_4 + 8H_2O \rightarrow 4H_3PO_2 + 4H^+ + 4e^-]$$
$$P_4 + 12H^+ + 12e^- \rightarrow 4PH_3$$

---

$$4P_4 + 24H_2O \rightarrow 12H_3PO_2 + 4PH_3$$

or

$$P_4 + 6H_2O \rightarrow 3H_2PO_2 + PH_3$$

Reaction 2,

$$H_3PO_2 + 2H_2O + 4AgNO_3 \rightarrow 4HNO_3 + H_3PO_4 + 4Ag$$

is a redox reaction:

$$H_3PO_2 + 2H_2O \rightarrow H_3PO_4 + 4H^+ + 4e^-$$
$$4[e^- + Ag^+ + NO_3^- \rightarrow Ag + NO_3^-]$$

---

$$H_3PO_2 + 2H_2O + 4Ag^+ + 4NO_3^- \rightarrow 4H^+ + H_3PO_4 + 4Ag + 4NO_3^-$$

or

$$H_3PO_2 + 2H_2O + 4AgNO_3 \rightarrow 4HNO_3 + H_3PO_4 + 4Ag$$

Reaction 3,

$$PH_3 + 3AgNO_3 \rightarrow 3HNO_3 + Ag_3P$$

is not a redox reaction.

When you balance any equation by the ion-electron method, you can check your answer by answering each of the following questions:

(a) Is the same number of atoms of each element on both sides of the equation?

(b) Is the total charge the same on each side of the equation?

(c) Do particular ions or molecules appear only once in the equation?

(d) Are the coefficients in the equation the lowest whole numbers?

If the answer to each question is *yes*, you have balanced the equation correctly.

(b) $P_4$ ox agent

$P_4$ red agent

Reaction (1) is a disproportionation reaction since $P_4$ is both the oxidizing and reducing agent and undergoes oxidation and reduction.

$Ag^+$ ox agent

$H_3PO_4$

    red agent

In reaction (2) $H_3PO_2$ is oxidized and is therefore the reducing agent. the silver ion, $Ag^+$, is reduced and is therefore the oxidizing agent.

Not a redox      Reaction (3) is of a type described in Section 9.14 of
reaction         your text.

15. [8.5]

(a) $2MnO_4^- + 5H_2C_2O_4 + 6H^+ \rightarrow 2Mn^{2+} + 10CO_2 + 8H_2O$

(b) $Sn^{2+} + 2HgCl_2 \quad Hg_2Cl_2 + Sn^{4+} + 2Cl^-$

(c) $8H_2O + 3MnO_4^- + 5Mo^{3+} \rightarrow 5MoO_4^{2-} + 3Mn^{2+} + 16H^+$

(d) $2Ce^{4+} + H_3AsO_3 + H_2O \rightarrow 2Ce^{3+} + H_3AsO_4 + 2H^+$

16. [8.5]

(a) $IO_4^- + 2I^- + H_2O \rightarrow IO_3^- + I_2 + 2OH^-$

(b) $CO(NH_2)_2 + 3OBr^- \rightarrow CO_2 + N_2 + 3Br^- + 3H_2O$

(c) $2OH^- + 2Al + 6H_2O \rightarrow 2Al(OH)_4^- + 3H_2$

## II. Oxygen

1.               The equation for the conversion of oxygen into ozone is

$$3O_2 \overset{h\nu}{\rightarrow} 3O_3$$

The oxidation number of an oxygen atom in $O_3$ is 0.

2.               The process described by the equation

$$Ca^{2+}(aq) + H_2O + CO_2(g) \rightarrow CaCO_3(s) + 2H^+(g)$$

ties up oxygen as a solid carbonate deposit. Respiration
consumes oxygen, but the oxygen can be released by photosynthesis.

3.               The equation for the atmospheric oxidation of FeO is

$$4FeO(s) + O_2(g) \rightarrow 2Fe_2O_3(s)$$

4.               The equations for the oxidations of $C(s)$ and $C_8H_{18}(l)$ are

$$C(s) + O_2(g) \rightarrow CO_2(g)$$

$$2C_8H_{18}(l) + 25O_2(g) \rightarrow 16CO_2(g) + 18H_2O(g)$$

5.               The equations for the oxidations of the volcanic gases are

$$2CO(g) + O_2(g) \rightarrow 2CO_2(g)$$

$$2SO_2(g) + O_2(g) \rightarrow 2SO_3(g)$$

$$2H_2(g) + O_2(g) \rightarrow 2H_2O(g)$$

## III. Hydrogen

1. $2.5 \times 10^{13}$    We calculate the number of tons of $H_2$ in the atmosphere
   tons $H_2$            of the earth:

$$? \text{ tons } H_2$$

$$= 5 \times 10^{15} \text{ tons atmosphere} \left( \frac{5 \text{ tons } H_2}{1000 \text{ tons atmosphere}} \right)$$

$$= 2.5 \times 10^{13} \text{ tons } H_2$$

2. $1.8 \times 10^{17}$ tons $H_2$    We calculate the number of tons of $H_2$ as follows:

$$? \text{ tons } H_2$$

$$= 1.5 \times 10^{24} \text{ g } H_2O \left( \frac{1 \text{ mol } H_2O}{18.0 \text{ g } H_2O} \right) \left( \frac{2.02 \text{ g } H_2}{1 \text{ mol } H_2O} \right) \left( \frac{1 \text{ lb } H_2}{4.54 \text{ g } H_2} \right) \left( \frac{1 \text{ ton } H_2}{2000 \text{ lb } H_2} \right)$$

$$= 1.8 \times 10^{17} \text{ tons } H_2$$

## SELF-TEST

Complete the test in 30 minutes:

1. Balance the following equations:

   (a) $ClO_2 \rightarrow ClO_2^- + ClO_3^-$ (basic solution)

   (b) $IO_4^- + H_2AsO_3^- \rightarrow IO_3^- + H_2AsO_4^-$ (acid solution)

   (c) $Sb + NO_3^- \rightarrow Sb_4O_6 + NO$ (acid solution)

   (d) $O_3 + I^- \rightarrow H_2O + I_2$

   (e) $NO_3^- + H_2S \rightarrow NO + S$ (acid solution)

   (f) $Cr_2O_7^{2-} + Cl^- \rightarrow Cr^3 + Cl_2$ (acid solution)

2. Determine the equivalent weight (to three significant figures) of each reactant in the equations in problem 1 above.

3. Complete the following:

   (a) $Mg(s) + H_2O(g) \rightarrow$

   (b) $H_2(g) + Cl_2(g) \rightarrow$

   (c) $Na(s) + H_2O(l) \rightarrow$

# Solutions

OBJECTIVES

I. You should be able to demonstrate your knowledge of the
following terms by defining them, describing them, or giving specific examples of them:

azeotrope [9.11]
colligative properties [9.10]
concentration [9.1, 9.6]
Debye-Hückel theory [9.13]
electrolyte [9.12]
equivalent weight [9.7]
entropy [9.4]
fractional distillation [9.11]
Gibbs free energy, $G$ [9.4]
infinite dilution [9.3]
lattice energy [9.4]
Le Chatelier's principle [9.5]
London forces [9.2]

metathesis reactions [9.14]
molality, *m* [9.6]
molarity, *M* [9.6]
mole fraction, *X* [9.6]
normality, *N* [9.6]
osmosis [9.10]
osmotic pressure [9.10]
Raoult's law [9.8]
saturation [9.1]
simple distillation [9.11]
solute [9.1]
solvent [9.1]
standard solution [9.7]
supersaturation [9.1]
titration [9.7]
van't Hoff factor [9.13]
weight percentage [9.6]

II. You should be able to determine the molarity, *M*, the molality, *m*, and the normality, *N*, of a solution. You should also be able to determine the weight percentage of solute in a solution and the mole fraction, *X*, of a component of a solution.

III. You should be able to work problems involving the preparation of dilute solutions from weighed samples or more concentrated solutions.

IV. You should be able to use normalities in computations involving oxidation—reduction and neutralization reactions.

V. You should be able to determine the vapor pressure of a component of an ideal solution.

VI. You should be able to determine molecular weights from data on freezing point depression and boiling point elevation.

VII. You should be able to write balanced metathesis equations in proper net ionic form. Memorize Tables 9.6 and 9.7 in your text.

EXERCISES        I. Work the following problems to gain experience in using
various expressions of concentration. Remember the equa
tions for molarity, $M$, normality, $N$, molality, $m$, and
mole fraction, $X$:

$$M = \frac{\text{number of moles of solute}}{\text{number of liters of solution}}$$

$$N = \frac{\text{number of gram equivalent weights of solute}}{\text{number of liters of solution}}$$

$$m = \frac{\text{number of moles of solute}}{\text{number of kilograms of solvent}}$$

$$X = \frac{\text{number of moles of a component}}{\text{total number of moles of all components}}$$

1. A total of 4.00 g of $AgNO_3$ is dissolved in a small
quantity of water and diluted to 1.00 liter. What is
the molarity of the final $AgNO_3$ solution?

2. The silver nitrate solution of problem 1 above is
to be used for titrating the chloride in natural
ground water:

$$Ag^+(aq) + Cl^-(aq) \rightarrow AgCl(s)$$

What is the normality of the silver nitrate solution?

3. A solution is $0.625M$ $KMnO_4$. What is the normality
of the solution if the permanganate is to be used
in the reaction

$$MnO_4^- + Fe^{2+} \rightarrow Mn^{2+} + Fe^{3+}$$

4. How many grams of $K_2Cr_2O_7$ are needed to prepare 1.00
liter of $0.100M$ $K_2Cr_2O_7$?

5. A beverage contains 0.174 g of ethanol, $C_2H_5OH$, and
4.72 g of water, $H_2O$. What is the mole fraction of
ethanol?

6. What is the molality of ethanol in the solution described in problem 5 above?

7. How many grams of $K_2Cr_2O_7$ are needed to prepare 60 ml
of a $0.15M$ $K_2Cr_2O_7$ solution?

II. Often it is necessary to prepare dilute solutions from
concentrated solutions by dilution. Picture the following dilution:

(a) A specified amount of solute from container A is
transferred to container B, which is empty:

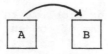

(b) The solute in container B is diluted by the addition
of solvent from container C:

The same amount of solute exists in container B *before*
and *after* dilution, but the amount of solvent is
changed. If we designate the condition before dilution by 1 and that after dilution by 2 and express the
amount of solute transferred in moles,

$$moles_1 = moles_2$$

and

$$M_1V_1 = M_2V_2$$

in which $M$ is molarity and $V$ is volume.

If we express the amount of solute transferred in
equivalents,

$$equivalents_1 = equivalents_2$$

and

$$N_1V_1 = N_2V_2$$

in which $N$ is normality and $V$ is volume.

We can also express the amount of solute transferred
in grams:

$$grams_1 = grams_2$$

and

$$(g/100 \text{ ml solvent})_1(V_1) = (g/100 \text{ ml solvent})_2(V_2)$$

Work the following problems:

1. What volume of a 0.45$M$ $KMnO_4$ solution is needed to
   prepare 1.0 liter of 0.10$M$ $KMnO_4$?

2. What volume of a 30.5% HCl solution, which has a
   density of 1.12 g/ml, is needed to prepare 500 ml
   of 0.200$M$ HCl?

3. What volume of a standard riboflavin solution that
   has a concentration of 1.00 mg/ml is necessary to prepare 100 ml of a solution that has a concentration
   of 0.100 mg/ml?

4. What is the molarity of an acetic acid solution if 25 ml of this solution are diluted to 100 ml to form a 0.75$M$ solution?

5. How many milliliters of 0.065$M$ riboflavin, vitamin $B_2$, are necessary to prepare 1.0 liter of riboflavin solution that contains a solute concentration of 0.10 mg/ml? The molecular weight of riboflavin is 376.

III. Work the following problems. The problems concern material presented in Chapter 9 and will serve as a review.

1. In which of the following liquids should the gas HF be most soluble: $CF_4$, $Br_2$, or $H_2O$?

2. Which of the following expressions of concentration is (are) independent of temperature: molarity, molality, or normality?

3. Which of the following solutions would have the lowest freezing point? 1.00$m$ $AlCl_3$, 1.00$m$ $CaCl_2$, or 1.00$m$ NaCl?

4. The solubility curve for $CoSO_4$ is shown in Figure 9.1 of the study guide. Refer to the curve and determine whether the solution process of $CoSO_4$ in $H_2O$ is endothermic or exothermic.

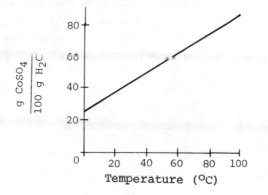

FIGURE 9.1   Solubility curve for $CoSO_4$

5. Is the absolute value of the lattice energy of $CoSO_4$ greater than than of the hydration energy? (See Figure 9.1.)

6. The aluminum compound $Al_2(SO_4)_3$ is a strong electrolyte. What is the vapor pressure above a solution prepared with 0.100 mol of $Al_2(SO_4)_3$ and 9.00 mol of $H_2O$ at 30°C if the vapor pressure of pure $H_2O$ at 30°C is 0.0395 atm?

7. Using Toepfer's reagent indicator, one can determine the concentration of free HCl present in gastric juice by reacting it with NaOH; usually concentrations are between 0 and 0.04$N$. If 5.0 ml of gastric juice requires 1.0 ml of 0.10$N$ NaOH for complete neutralization, determine whether a patient's level of free HCl is within the normal limits.

8. The heating of 400 ml of $BaCl_2$ solution, which has a density of 1.09 g/ml, produces 43.7 g of dry, solid residue. What is the molarity of the original $BaCl_2$ solution? What is the molality?

9. Is Henry's law valid for all solutions?

10. If NaCl costs $56.00 per 1.0 ton and $CaCl_2$ costs $88.00 per 1.0 ton, which compound would be more economical to prevent icy roads?

11. According to Raoult's law, the lowering of vapor pressure of the solvent in dilute solutions of non-volatile soluts is proportional to what?

12. Does the entropy of a natural process increase or decrease?

13. Which of the following would be expected to have the greater hydration energy: $BeCl_2$ or $BaCl_2$?

14. At 25°C the osmotic pressure of 100 ml of β-lacto-globulin solution containing 1.49 g of that protein was found to be 0.0100 atm. Calculate the molecular weight of the protein.

15. The vapor pressure of pure water at 25°C is $3.12 \times 10^{-2}$ atm. What is the vapor pressure of a 100-g sample of water in which 27.4 g of sucrose, $C_{12}H_{22}O_{11}$, is dissolved?

16. Bacteria responsible for gangrene, *Clostridium welchii*, are anaerobic organisms; i.e., their growth is inhibited by molecular oxygen. Why are patients with gangrene sometimes placed in a high-pressure, oxygen-rich chamber (a hyperbaric chamber)?

17. Water boils at 95°C in Denver, Colorado. Why? Estimate how much NaCl should be added to 1000 g of $H_2O$ to raise the boiling point of water in Denver to 100°C.

18. Write balanced net ionic chemical equations for the reactions that occur when aqueous solutions of the following compounds are mixed. Some of them yield no reaction; indicate this by writing NR.

(a) $HCl + NaCl \rightarrow$

(b) $HCl + NH_4Cl \rightarrow$

(c) $HCl + AgNO_3 \rightarrow$

(d) $HCl + Na_2CO_3 \rightarrow$

(e) $HCl + NaOH \rightarrow$

(f) $HCl + NaC_2H_3O_2 \rightarrow$

(g) $HCl + HNO_3 \rightarrow$

(h) $NaOH + H_3PO_4 \rightarrow$

(i) $NaOH + K_2SO_4 \rightarrow$

(j) $NaOH + (NH_4)_2SO_4 \rightarrow$

(k) $NaOH + FeSO_4 \rightarrow$

(l) $(NH_4)_2CO_3 + Na_2SO_4 \rightarrow$

## ANSWERS TO EXERCISES

I. Expressions of concentration

**1. 0.0235$M$**
**[9.6]**

We calculate the molarity:

$$? \, M \, AgNO_3 = \frac{\text{number of moles of } AgNO_3}{\text{number of liters of } AgNO_3 \text{ soln.}}$$

$$= \frac{(4.00 \text{ g } AgNO_3)\left(\dfrac{1 \text{ mol } AgNO_3}{169.9 \text{ g } AgNO_3}\right)}{1 \text{ liter } AgNO_3 \text{ soln.}}$$

$$= 0.0235M \, AgNO_3$$

**2. 0.0235$N$**
**[9.6]**

For the reaction

$$Ag^+(aq) + Cl^-(aq) \rightarrow AgCl(s)$$

the equivalent weight of $AgNO_3$ is equal to the molecular weight of $AgNO_3$. Therefore,

$$N = M = 0.0235$$

**3. 3.12$N$**
**[9.6]**

For the reaction

$$5e^- + 8H^+ + \overset{7+}{MnO_4^-} \rightarrow Mn^{2+} + 4H_2O$$

the equivalent weight of $KMnO_4$ is 1/5 of the molecular weight of $KMnO_4$. Thus, the normality is 5 times the molarity since 1 mol $KMnO_4$ is 5 equivalents:

$$? \ N \ KMnO_4 = \left(\frac{0.625 \text{ mol } KMnO_4}{1 \text{ liter } KMnO_4 \text{ soln.}}\right)\left(\frac{5 \text{ equivalents } KMnO_4}{1 \text{ mol } KMnO_4}\right)$$

$$= \left(\frac{3.12 \text{ equivalents } KMnO_4}{1 \text{ liter } KMnO_4 \text{ soln.}}\right)$$

$$= 3.12N \ KMnO_4$$

4. 29.4 g
[9.6]

We calculate the number of grams of $K_2Cr_2O_7$ as follows:

$$M \ K_2Cr_2O_7 = \frac{\text{number of moles of } K_2Cr_2O_7}{\text{number of liters of } K_2Cr_2O_7 \text{ soln.}}$$

$$0.1.00M \ K_2Cr_2O_7 = \frac{? \text{ g } K_2Cr_2O_7\left(\frac{1 \text{ mol } K_2Cr_2O_7}{294.2 \text{ g } K_2Cr_2O_7}\right)}{1 \text{ liter } K_2Cr_2O_7 \text{ soln.}}$$

$$? \text{ g } K_2Cr_2O_7$$

$$= \frac{(0.100M \ K_2Cr_2O_7)(1 \text{ liter } K_2Cr_2O_7 \text{ soln.})(294.2 \text{ g } K_2Cr_2O_7)}{1 \text{ mol } K_2Cr_2O_7}$$

$$= 29.4 \text{ g } K_2Cr_2O_7$$

5. 0.0142
[9.6]

We calculate the mole fraction of ethanol as follows:

$$? \text{ mol } C_2H_5OH = 0.174 \text{ g } C_2H_5OH\left(\frac{1 \text{ mol } C_2H_5OH}{46.07 \text{ g } C_2H_5OH}\right)$$

$$= 0.003776 \text{ mol } C_2H_5OH$$

$$? \text{ mol } H_2O = 4.72 \text{ g } H_2O\left(\frac{1 \text{ mol } H_2O}{18.02 \text{ g } H_2O}\right)$$

$$= 0.2619 \text{ mol } H_2O$$

$$X \ C_2H_5OH = \frac{\text{number of moles of } C_2H_5OH}{\text{total number of moles of } C_2H_5OH + H_2O}$$

$$= \frac{0.003776 \text{ mol}}{0.003776 \text{ mol} + 0.2619 \text{ mol}} = 0.0142$$

6. 0.800m
[9.6]

We calculate the molality of ethanol as follows:

$$? \ m \ C_2H_5OH = \frac{\text{number of moles of } C_2H_5OH}{\text{number of kilograms of } H_2O}$$

$$= \frac{0.003776 \text{ mol } C_2H_5OH}{(4.72 \text{ g } H_2O)\left(\frac{1 \text{ kg}}{10^3 \text{ g}}\right)}$$

$$= 0.800m \ C_2H_5OH$$

7. 2.6 g          Rearranging the equation
   [9.6]

$$M\ K_2Cr_2O_7 = \frac{\text{number of moles of } K_2Cr_2O_7}{\text{number of liters of } K_2Cr_2O_7 \text{ soln.}}$$

we find

($M\ K_2Cr_2O_7$) (number of liters of $K_2Cr_2O_7$ soln.)

= number of moles of $K_2Cr_2O_7$

$$= \frac{\text{number of grams of } K_2Cr_2O_7}{\text{molecular weight of } K_2Cr_2O_7}$$

Substituting values into the rearranged equation, we find

$$(0.15M\ K_2Cr_2O_7)\,(0.060 \text{ liter } K_2Cr_2O_7 \text{ soln.}) = \frac{?\ g\ K_2Cr_2O_7}{294\ g\ K_2Cr_2O_7}$$

$$?\ g\ K_2Cr_2O_7 = 2.6\ g\ K_2Cr_2O_7$$

We could also use a logic chain to solve the problem:

$?\ g\ K_2Cr_2O_7$

$= (0.15M\ K_2Cr_2O_7)\,(0.060 \text{ liter } K_2Cr_2O_7 \text{ soln.})\left(\dfrac{294\ g\ K_2Cr_2O_7}{1\ \text{mol } K_2Cr_2O_7}\right)$

$= 2.6\ g\ K_2Cr_2O_7$

## II. Dilution

1. 0.22 liter     The number of moles of solute in the unknown volume of
   [9.6]          concentrated solution equals the number of moles of
                  solute in the dilute solution:

$$V_1V_1 = M_2M_2$$

Rearranging the preceding equation and substituting
values into it, we find

$$V_1 = \frac{V_2M_2}{M_1}$$

$$= \frac{(1.0 \text{ liter})(0.10M)}{0.45M}$$

$$= 0.22 \text{ liter}$$

2. 10.7 ml        If we wish to use
   [9.6]
$$V_1V_1 = M_2M_2$$

and

($M\ K_2Cr_2O_7$) (number of liters of $K_2Cr_2O_7$ soln.)

$$= \frac{\text{number of grams of } K_2Cr_2O_7}{\text{molecular weight of } K_2Cr_2O_7}$$

Substituting values into the rearranged equation, we find

$$(0.15 \quad K_2Cr_2O_7)(0.060 \text{ liter } K_2Cr_2O_7 \text{ soln.})$$

$$= \frac{? \text{ g } K_2Cr_2O_7}{294 \text{ g } K_2Cr_2O_7}$$

$$? \text{ g } K_2Cr_2O_7 = 2.6 \text{ g } K_2Cr_2O_7$$

3. 10.0 ml
   [9.6]

In general,

$$V_1C_1 = V_2C_2$$

in which $V$ is volume and $C$ is concentration. In this problem the concentrations are expressed in mg/ml. If we rearrange the preceding equation and substitute values into it, we find

$$V_1 = \frac{V_2V_2}{C_1}$$

$$= \frac{(100 \text{ ml})(0.100 \text{ mg/ml})}{(1.00 \text{ mg/ml})} = 10.0 \text{ ml}$$

4. 3.0$M$
   [9.6]

Rearranging the equation

$$V_1M_1 = V_2M_2$$

and substituting values into it, we find

$$M_1 = \frac{V_2M_2}{V_1}$$

$$? \; M_1 = \frac{(100 \text{ ml})(0.75 \; )}{(25 \text{ ml})}$$

$$= 3.0M$$

5. 4.1 ml

First we change one of the concentrations so that it has the same units as the other. We determine the molarity of the 0.10 mg/ml solution of vitamin $B_2$:

$$? \; M \; B_2 = \left(\frac{0.10 \text{ mg } B_2}{1 \text{ ml soln.}}\right)\left(\frac{1 \text{ g}}{1000 \text{ mg}}\right)\left(\frac{1000 \text{ ml}}{1 \text{ liter}}\right)\left(\frac{1 \text{ mol } B_2}{376 \text{ g } B_2}\right)$$

$$= 2.66 \times 10^{-4}M \; B_2$$

Then we rearrange the equation

$$V_1M_1 = V_2M_2$$

and substitute values into it:

$$V_1 = \frac{V_2M_2}{M_1}$$

$$V_1 = \frac{(1.0 \text{ liter})(2.66 \times 10^{-4} \quad B_2)}{6.5 \times 10^{-2}M \; B_2}$$

$$= 0.0041 \text{ liter, or } 4.1 \text{ ml}$$

### III. Properties of solutions

**1.** $H_2O$
[9.2]

*Like dissolves like*. Hydrogen fluoride, HF, and water, $H_2O$, are highly polar and have hydrogen bonding ability. Tetrafluoromethane, $CF_4$, and bromine, $Br_2$, are nonpolar.

**2.** molality
[9.6]

Molality is defined as the number of moles of solute per kilogram of solvent:

$$m = \frac{\text{number of moles of solute}}{\text{number of kilograms of solvent}}$$

Molality is not defined in terms of volume, as are molarity and normality:

$$M = \frac{\text{number of moles of solute}}{\text{number of liters of solution}}$$

$$N = \frac{\text{number of equivalents of solute}}{\text{number of liters of solution}}$$

Volume changes with a change in temperature. In problems concerning boiling point elevation and freezing point depression, $m$ is used instead of $M$ or $N$.

**3.** $1.00m$ $AlCl_3$
[9.9, 9.13]

If we assume each compound to be a strong electrolyte, each molecule of compound would dissociate into the following number of ions:

$AlCl_3$: 4
$CaCl_2$: 3
NaCl: 2

If we substitute values into the equation

$$\Delta T_f = iK_f m$$

we find that the freezing point depression is largest for $AlCl_3$, for which $i = 4$.

**4.** endothermic
[9.5]

As the temperature increases, the amount of $CoSO_4$ that dissolves increases:

$$H_2O + CoSO_4 \rightarrow Co^{2+}(aq) + SO_4^{2-}(aq)$$

The equilibrium shifts therefore to the right. According to Le Chatelier's principle, the application of stress to a system in equilibrium causes the system to react to counteract the stress, thus establishing a new equilibrium state. If a temperature rise (the addition of heat to the system) causes the equilibrium to shift to the right, heat must appear on the left side of the equation:

$$\text{heat} + H_2O + CoSO_4 \rightarrow Co^{2+}(aq) + SO_4^{2-}(aq)$$

If heat is required in the solution process, the process is endothermic.

5. yes
   [9.4]

When a substance dissolves, energy (the lattice energy) is required to break the crystal lattice, and energy (the hydration energy) is released by the hydration of the ions. The solution process for $CoSO_4$ is endothermic; i.e., in the process more energy is required than is released. Thus, the absolute value of the lattice energy is greater than that of the hydration energy.

6. 0.0374 atm
   [9.8]

Since $Al_2(SO_4)_3$ is a strong electrolyte, in solution a mole of it produces 5 mol of ions:

$$Al_2(SO_4)_3 \rightarrow 2Al^{3+} + 3SO_4^{2-}$$

Assuming an ideal solution is formed, we can use Raoult's law to solve the problem:

$$p_{H2O} = X_{H2O}p_{H2O}^{o}$$

$$= 0.0395 \text{ atm}\left(\frac{9.00 \text{ mol}}{9.00 \text{ mol} + 0.500 \text{ mol}}\right)$$

$$= 0.0374 \text{ atm}$$

Since the solute is nonvolatile,

$$P_{total} = 0.0374 \text{ atm}$$

7. yes, 0.020N
   [9.7]

If we substitute values into the equation

$$V_1N_1 = V_2N_2$$

or into its rearranged form

$$N_1 = \frac{V_2N_2}{V_1}$$

we find

$$? N = \frac{(1.0 \text{ ml})(0.10N)}{(5.0 \text{ ml})}$$

$$= 0.020N$$

8. 0.525M
   0.535m
   [9.6]

When the heating of the solution is complete, 43.7 g of $BaCl_2$ remain as residue. Thus, we can calculate the molarity:

$$? M \text{ BaCl}_2 = \frac{\text{number of moles of BaCl}_2}{\text{number of liters of BaCl}_2 \text{ soln.}}$$

$$= \frac{(43.7 \text{ g BaCl}_2)\left(\frac{1 \text{ mol BaCl}_2}{208.2 \text{ g BaCl}_2}\right)}{(400 \text{ ml BaCl}_2 \text{ soln.})\left(\frac{1 \text{ liter BaCl}_2 \text{ soln.}}{1000 \text{ ml BaCl}_2 \text{ soln.}}\right)}$$

$$= 0.525M \text{ BaCl}_2$$

Determining the molality is not as straightforward as determining the molarity. Since kg solvent, not kg solution, is needed, we must make use of the density and the fact that the weight of the solution equals the weight of water plus the weight of $BaCl_2$. We first calculate the weight of the solution from the density:

$$? \text{ g soln.} = 400 \text{ ml soln.} \left(\frac{1.09 \text{ g soln.}}{1 \text{ mol soln.}}\right) = 436 \text{ g soln.}$$

Of the 436 g of solution, we know that 43.7 g are $BaCl_2$. Thus, we can calculate the weight of water:

$$? \text{ g } H_2O = 436 \text{ g soln.} - 43.7 \text{ g } BaCl_2 = 392 \text{ g } H_2O$$

We can now calculate the molality of $BaCl_2$:

$$? \, m \, BaCl_2 = \frac{\text{number of moles of } BaCl_2}{\text{number of kg of water}}$$

$$= \frac{(43.7 \text{ g } BaCl_2)\left(\dfrac{1 \text{ mol } BaCl_2}{208.2 \text{ g } BaCl_2}\right)}{(392 \text{ g } H_2O)\left(\dfrac{1 \text{ kg } H_2O}{1000 \text{ kg } H_2O}\right)}$$

$$= 0.535m \, BaCl_2$$

9. no
[9.5]

Henry's law, which states that the solubility of a gas in a solution is directly proportional to the partial pressure of that gas above the solution, is valid only for dilute solutions and low pressures. Also, extremely soluble gases usually react with the solvent and thus do not follow Henry's law.

10. NaCl
[9.9, 9.13]

We can calculate the cost of a single mole of ions for each compound:

For $CaCl_2$:

$$? \text{ \$/mol ions}$$

$$= \left(\frac{\$88}{2000 \text{ lb } CaCl_2}\right)\left(\frac{1 \text{ lb } CaCl_2}{454 \text{ g } CaCl_2}\right)\left(\frac{111 \text{ g } CaCl_2}{1 \text{ mol } CaCl_2}\right)\left(\frac{1 \text{ mol } CaCl_2}{3 \text{ mol ions}}\right)$$

$$= 3.6 \times 10^{-3} \text{ \$/mol ions}$$

For NaCl:

$$? \text{ \$/mol ions}$$

$$= \left(\frac{\$56}{2000 \text{ lb NaCl}}\right)\left(\frac{1 \text{ lb NaCl}}{454 \text{ g NaCl}}\right)\left(\frac{58.4 \text{ g NaCl}}{1 \text{ mol NaCl}}\right)\left(\frac{1 \text{ mol NaCl}}{2 \text{ mol ions}}\right)$$

$$= 1.8 \times 10^{-3} \text{ \$/mol ions}$$

Thus, it is cheaper to use NaCl than $CaCl_2$.

11. Mole fraction
    of solute par-
    ticles [9.8]

12. increase
    [9.4]

13. $BeCl_2$
    [9.2]

An ion with a larger value of the ratio of ionic charge to ionic radius forms a more stable hydrated ion and thus has a larger enthalpy of hydration, $\Delta H_{hydration}$.

14. $3.64 \times 10^4$
    g/mol
    [9.10]

Rearranging the van't Hoff equation

$$\pi V = nRT$$

we find

$$n = \frac{\pi V}{RT}$$

Since

$$n = \text{number of moles} = \frac{\text{weight in grams}}{\text{molecular weight in grams}}$$

the van't Hoff equation becomes

$$\frac{\text{weight in grams}}{\text{molecular weight in grams}} = \frac{\pi V}{RT}$$

or

$$\text{molecular weight in grams} = \frac{(RT)(\text{weight in grams})}{\pi V}$$

Substituting values into the van't Hoff equation, we find

molecular weight in grams

$$= \frac{(0.08206 \text{ liter atm mol}^{-1} \text{ K}^{-1})(298°\text{K})(1.49 \text{ g})}{(1.00 \times 10^{-2} \text{ atm})(1.00 \times 10^2 \text{ ml})(1 \text{ liter}/10^3 \text{ ml})}$$

$$= 3.64 \times 10^4 \text{ g/mol}$$

15. $3.08 \times 10^{-2}$ atm
    [9.8]

We can use Raoult's law to solve this problem:

$$p_{H_2O} = X_{H_2O} p^o_{H_2O}$$

We must calculate the mole fraction of $H_2O$. First, however, we must determine the number of moles of water and sucrose:

$$? \text{ mol H}_2\text{O} = 100 \text{ g H}_2\text{O}\left(\frac{1 \text{ mol H}_2\text{O}}{18.02 \text{ g H}_2\text{O}}\right)$$

$$= 5.55 \text{ mol H}_2\text{O}$$

$$? \text{ mol C}_{12}\text{H}_{22}\text{O}_{11} = 27.4 \text{ g C}_{12}\text{H}_{22}\text{O}_{11}\left(\frac{1 \text{ mol C}_{12}\text{H}_{22}\text{O}_{11}}{342.3 \text{ g C}_{12}\text{H}_{22}\text{O}_{11}}\right)$$

$$= 8.00 \times 10^{-2} \text{ mol C}_{12}\text{H}_{22}\text{O}_{11}$$

The mole fraction of water is therefore

$$x_{H_2O} = \frac{5.55 \text{ mol}}{5.55 \text{ mol} + 0.800 \text{ mol}}$$

$$= 0.986$$

Substituting values into Raoult's law, we find

$$p_{H_2O} = x_{H_2O}p_{H_2O}^{o}$$

$$= 0.986(3.12 \times 10^{-2} \text{ atm})$$

$$= 3.08 \times 10^{-2} \text{ atm}$$

Since the solution is nonvolatile,

$$p_{total} = p_{H_2O} = 3.08 \times 10^{-2} \text{ atm}$$

16. [9.5]

According to Henry's law, the amount of gas dissolved in solution increases as the partial pressure of the gas increases. Therefore, since the pressure of oxygen in a hyperbaric chamber is greater than that in the normal atmosphere, more oxygen will dissolve in the body fluids when a patient is in a hyperbaric chamber than when in the normal atmosphere. The greater the amount of oxygen in the body fluids, the greater the inhibition of growth of *Clostridium welchii*.

17. $2.9 \times 10^2$
g NaCl
[7.4, 9.9]

Atmospheric pressure in Denver is lower than 1 atmosphere; thus, the boiling point of water in Denver is lower than the normal boiling point, i.e., the temperature at which the vapor pressure of water equals 1 atmosphere.

We must calculate the amount of NaCl necessary to raise the boiling point of 1000 g of $H_2O$ to 100°C in Denver. We can calculate the molality of the solution from the equation for boiling point elevation:

$$\Delta T_b = imK_b$$

Rearranging the previous equation and substituting values into it, we find

$$? \ m = \frac{\Delta T_b}{iK_b}$$

$$= \frac{(100 - 95)°C}{2(0.512°C/m)}$$

$$= 4.9m$$

We then calculate the number of grams of NaCl necessary for the 1000 g $H_2O$:

$$? \ \text{g NaCl} = \left(\frac{5 \text{ mol NaCl}}{1000 \text{ g } H_2O}\right)(1000 \text{ g } H_2O)\left(\frac{58.4 \text{ g NaCl}}{1 \text{ mol NaCl}}\right)$$

$$= 2.9 \times 10^2 \text{ g NaCl}$$

The NaCl solution is very concentrated, and it is doubtful that NaCl is completely dissociated; i.e., $i$ is less
than 2. Although the concentration of NaCl necessary to
increase the boiling point by 5°C would be larger than
4.9$m$, we have made a good first approximation.

18. [9.14]
    (a) NR
    (b) NR
    (c) $Cl^- + Ag^+ \rightarrow AgCl(s)$
    (d) $2H^+ + CO_3^{2-} \rightarrow CO_2(g) + H_2O$
    (e) $H^+ + OH^- \rightarrow H_2O$
    (f) $H^+ + C_2H_3O_2^- \rightarrow HC_2H_3O_2$
    (g) NR
    (h) $3OH^- + H_3PO_4 \rightarrow 3H_2O + PO_4^{3-}$
    (i) NR
    (j) $OH^- + NH_4^+ \rightarrow NH_3(g) + H_2O$
    (k) $2OH^- + Fe^{2+} \rightarrow Fe(OH)_2(s)$
    (l) NR

## SELF-TEST

I. Complete the test in 30 minutes:

1. A 0.2070-g sample of an unknown organic acid is dissolved in 100 ml of 0.02560$N$ sodium hydroxide. The
   resulting solution is titrated to the end point with
   36.50 ml of 0.0240$N$ NCl. Calculate the equivalent
   weight of the unknown acid.

2. A 20.00-ml portion of commercial vinegar, the density
   of which is 1.01 g/ml, is titrated with 28.75 ml of
   0.503$M$ NaOH. What is the weight percentage of acid,
   $HC_2H_3O_2$, in vinegar?

3. A sample of sodium is added to water to make 500 ml of
   solution. At STP 33.6 liters of dry hydrogen gas are
   collected from the reaction of sodium metal with water:

   $2Na(s) + 2H_2O \rightarrow 2NaOH(aq) + H_2(g)$

   What is the normality of the NaOH solution produced
   by the reaction?

4. What volume of a concentrated hydrochloric acid solution, which is 37% HCl and has a specific gravity of
   1.189 g/ml, should be used to prepare 2000 ml of
   0.100$M$ HCl solution? Specific gravity is the ratio of
   the density of a material to the density of water.

5. A solution containing 0.500 g of an unknown nonvolatile solute in 10.0 g of camphor has a freezing point
   of 159.4°C. What is the molecular weight of the solute if the normal freezing point of camphor is 179.0°C
   and the molal freezing-point depression constant for
   camphor is 49°C/$m$?

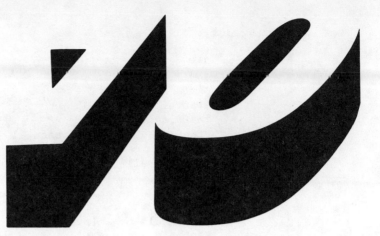

# Electrochemistry

OBJECTIVES    I. You should be able to demonstrate your knowledge of the
              following terms by defining them, describing them, or giv-
              ing specific examples of them:

                 ampere [10.1]
                 anode [10.2]
                 cathode [10.2]
                 concentration polarization [10.11]
                 conduction [10.1, 10.2]
                 coulomb [10.1]
                 current [10.1]
                 disproportionation [10.7]
                 electrolysis [10.3]
                 electrolytic cell [10.2]
                 electromotive force, emf [10.6]
                 faraday [10.4]
                 Faraday's laws [10.4]
                 fuel cell [10.12]

Nernst equation [10.8, 10.9]
ohm [10.1]
Ohm's law [10.1]
overvoltage [10.10]
resistance [10.1]
standard electrode potential [10.7]
volt [10.1]
voltaic cell [10.5]

II. You should be able to use Faraday's laws.

III. You should be able to determine standard electrode potentials for reactions from tabulated values.

IV. Using a table of standard electrode potentials, you should be able to determine the potential of an electrochemical cell for any set of concentrations of reactants and products.

V. You should be able to diagram electrochemical cells and determine the electrode signs, the direction of the electron movement, and the direction of ion migration.

UNITS,
SYMBOLS,
MATHEMATICS

I. You should understand and be able to manipulate logarithms in order to work problems using the Nernst equation. Please read the appendix in your text.

Do not depend on your calculator entirely. There are limits to its ability, but it will cause you no difficulty if you understand what you are doing. Common logarithms are powers, i.e., exponents, to which the number 10 must be raised to obtain the number. For example:

$$100 = 10^2$$

so

$$\log 100 = 2$$

or 10 must be raised to the second power to obtain the number 100. Also remember that

$$1 = 10^0$$

$$\log 1 = 0$$

and

$$10 = 10^1$$

$$\log 10 = 1$$

To avoid difficulty in determining the logarithm of a
number other than 1, 10, 100, and so on, you should
always write the number in proper scientific notation.
(See Chapter 1 of the study guide.) When this is done,
you will have a number between 1 and 10, which is multiplied by 10 raised to a whole number exponent. For
example, the number 106 is written as

$$1.06 \times 10^2$$

Then take the log of the number written in scientific
notation:

$$\log (1.06 \times 10^2) = \log 1.06 + \log 10^2$$

Remember that the log of a number between 1 and 10 is
a number between 0 and 1, so

$$\log 1.06 + \log 10^2 = 0.025 + 2 = 2.025$$

There are only three significant figures in 2.025. The
number before the decimal, called the characteristic,
indicates the power to which 10 is raised when the original number is written in proper scientific notation,
and the decimal portion, called the mantissa, gives the
rest of the number. Each number, including the zero in a
mantissa, is significant. Therefore, the following logarithms are reported with three significant figures:

$$\log (1.06 \times 10^2) = 2.025$$

$$\log (1.06 \times 10^{24}) = 24.025$$

$$\log (1.06 \times 10^{124}) = 124.025$$

The same rules apply for small numbers. Three significant
figures are reported here:

$$\log (1.06 \times 10^{-24}) = 0.025 - 24 = 23.975$$

$$\log (1.06 \times 10^{-124}) = 123.975$$

Do the following on your calculator, but make sure that
you can report the result properly:

$$\log 76.23 = 1.8821$$

$$\log \left(\frac{7.623}{22.4}\right) = -0.468$$

$$\log (6.023 \times 10^{24}) = 24.7798$$

$$\log \left(\frac{6.023 \times 10^{24}}{1.7 \times 10^{-86}}\right) = 110.55$$

$$\log 7674 = \log (7.674 \times 10^3) = 3.8850$$

$$\log (0.0243) = -1.614$$
$$\log (1.0243) = 0.01043$$

EXERCISES

I. Answer each of the following with *true* or *false*. If a statement is false, correct it.

_____ 1. In an electrolyric cell, oxidation occurs at the positive electrode.

_____ 2. In a galvanic cell, oxidation occurs at the positive electrode.

_____ 3. The anode of a cell is indicated on the left in the cell notation.

_____ 4. In both galvanic and electrolytic cells, reduction occurs at the cathode.

_____ 5. A positive cell potential indicates that the reaction should occur spontaneously as written.

_____ 6. A battery is an electrolytic cell.

_____ 7. One faraday is required to reduce 1 mole of $Cu^{2+}$ ions to $Cu(s)$.

_____ 8. An Avogadro's number of $Ni^{2+}$ ions can be reduced to $Ni(s)$ with 2 faradays.

_____ 9. Four faradays produce 22.4 liters of $O_2$ at STP in the electrolysis of water:

$$2H_2O(l) \rightarrow O_2(g) + 4H^+(aq) + 4e^-$$

_____ 10. Electrode potentials are temperature dependent.

_____ 11. The best oxidizing agent in Table 10.1 (page 161 of the study guide) is $F_2$.

_____ 12. The best reducing agent in Table 10.1 is $Li^+$.

_____ 13. According to Ohm's law, potential difference is directly proportional to current, $I$.

_____ 14. In the electrolysis of a unit activity solution of $Na_2SO_4$ using Pt electrodes, the reaction occurring at the anode is

$$2H_2O(l) \rightarrow O_2(g) + 4H^+(aq) + 4e^-$$

_____ 15. According to Faraday's laws, 1 F would liberate 2 g of $H_2$ in the reduction of hydrogen ion.

_____ 16. The standard hydrogen electrode consists of hydrogen gas at 1 atm pressure bubbling over a Pt electrode that is immersed in an acid solution containing $H^+(aq)$ at unit activity.

**TABLE 10.1    Standard Electrode Potentials at 25°C**

| Half Reaction | $\mathscr{E}^\circ$ (volts) |
|---|---|
| $Li^+ + e^- \rightleftharpoons Li$ | -3.045 |
| $K^+ + e^- \rightleftharpoons K$ | -2.925 |
| $Ba^{2+} + 2e^- \rightleftharpoons Ba$ | -2.906 |
| $Ca^{2+} + 2e^- \rightleftharpoons Ca$ | -2.866 |
| $Na^+ + e^- \rightleftharpoons Na$ | -2.714 |
| $Mg^{2+} + 2e^- \rightleftharpoons Mg$ | -2.363 |
| $Al^{3+} + 3e^- \rightleftharpoons Al$ | -1.662 |
| $2H_2O + 2e^- \rightleftharpoons H_2 + 2OH^-$ | -0.82806 |
| $Zn^{2+} + 2e^- \rightleftharpoons Zn$ | -0.7628 |
| $Cr^{3+} + 3e^- \rightleftharpoons Cr$ | -0.744 |
| $Fe^{2+} + 2e^- \rightleftharpoons Fe$ | -0.4402 |
| $Cd^{2+} + 2e^- \rightleftharpoons Cd$ | -0.4029 |
| $Ni^{2+} + 2e^- \rightleftharpoons Ni$ | -0.250 |
| $Sn^{2+} + 2e^- \rightleftharpoons Sn$ | -0.136 |
| $Pb^{2+} + 2e^- \rightleftharpoons Pb$ | -0.126 |
| $2H^+ + 2e^- \rightleftharpoons H_2$ | 0 |
| $Cu^{2+} + 2e^- \rightleftharpoons Cu$ | +0.337 |
| $Cu^+ + e^- \rightleftharpoons Cu$ | +0.521 |
| $I_2 + 2e^- \rightleftharpoons 2I^-$ | +0.5355 |
| $Fe^{3+} + e^- \rightleftharpoons Fe^{2+}$ | +0.771 |
| $Ag^+ + e^- \rightleftharpoons Ag$ | +0.7991 |
| $Br_2 + 2e^- \rightleftharpoons 2Br^-$ | +1.0652 |
| $O_2 + 4H^+ + 4e^- \rightleftharpoons 2H_2O$ | +1.229 |
| $Cr_2O_7^{2-} + 14H^+ + 6e^- \rightleftharpoons 2Cr^3 + 7H_2O$ | +1.33 |
| $Cl_2 + 2e^- \rightleftharpoons 2Cl^-$ | +1.3595 |
| $MnO_4^- + 8H^+ + 5e^- \rightleftharpoons Mn^{2+} + 4H_2O$ | +1.51 |
| $F_2 + 2e^- \rightleftharpoons 2F$ | +2.07 |

17. The addition of the standard electrode potentials of the half reactions gives the value of the emf of a cell.

18. In the cell

    $Cd|Cd^{2+}||Ni^{2+}|Ni$

    the Cd electrode gains weight.

19. For the cell

    $Cr|Cr^{3+}(0.001M)||Fe^{2+}(0.01M)|Fe$

    the Nernst equation is

    $$\mathscr{E} = +0.304 \text{ V} - (0.0592/6) \log\left(\frac{(10^{-3})^2}{(10^{-2})^3}\right)$$

20. Lower voltage than that calculated is required for electrolysis if concentration polarization occurs.

———————————

21. An ampere is a coulomb/sec.

———————————

22. A faraday equals 0.0592 coulomb.

II. Answer each of the following:

1. Using data from Table 10.1 in the study guide, calculate the standard electrode potential for the half reaction $Cr_2O_7^{2-} \rightarrow Cr$.

2. Corrosion is an oxidation process. Often iron is coated with other metals to protect it from rusting. Determine what would happen to iron if it were coated with tin, as in tin cans, and the coating cracked, exposing the iron to air and water. (Hint: Assume that the metal forms an electrochemical cell in solution, the cathode reaction being the reduction of $O_2$ to $OH^-$.)

3. If the iron of problem 2 above were galvanized, i.e., coated with zinc, what would happen if the coating cracked and the iron became exposed to air and water. (Hint: Assume that the metal forms an electrochemical cell in solution, the cathode reaction being the reduction of $O_2$ to $OH^-$.)

4. During cellular respiration, oxidation-reduction reactions take place. If the free energy of the reaction NAD to FAD (nicotinamide adenine dinucleotide to flavin adenine dinucleotide) is -12.00 kcal/mol and the standard emf is +0.26 V, how many electrons are involved in this oxidation step in the respiratory chain?

5. In the Dow process for obtaining magnesium, the magnesium is precipitated from seawater as the hydroxide, which is dissolved in HCl after purification. The resulting magnesium chloride solution is evaporated, and the magnesium chloride is melted and electrolyzed. Answer the following:
   (a) How many hours would it take to obtain 5.0 kg of magnesium metal using $1.0 \times 10^4$ amperes?
   (b) What volume of chlorine gas at STP would be evolved in the time it takes to obtain the 5.0 kg of Mg metal using $1.0 \times 10^4$ amperes?

6. Predict whether or not each of the following reactions will occur spontaneously and write a balanced chemical equation for each reaction that is predicted to occur. Assume that each reactant and product is present at unit activity in aqueous solution at 25°C.

(a) $Cd^{2+}$ ion reduced to Cd by Ni
(b) Sn oxidized to $Sn^{2+}$ by $Cu^{2+}$ ion
(c) $I_2$ reduced to $I^-$ ion by Ag
(d) $MnO_4^-$ ion reduced to $Mn^{2+}$ ion by $Br^-$ ion

7. Write the shorthand cell notation for the spontaneous reactions in problem 6 above.

8. What is the potential of the cell?

$Pt|Fe^{3+}(0.010M), Fe^{2+}(0.12M)||Fe^{3+}(0.13M), Fe^{2+}(0.71M)|Pt$

9. Use Table 10.1 of the study guide to predict what will happen when a bar of cadmium is added to a solution of $Fe^{3+}$.

10. The salt $Co_2(SO_4)_3 \cdot 18H_2O$ decomposes when added to water. Why?

11. If 17.6 liters of $O_2$ at STP are evolved at the anode in an electrochemical cell and the only reaction at the cathode is the reduction of $Cu^{2+}$ to Cu(s), how many grams of Cu are formed?

12. Use the standard electrode potentials of bromine-containing compounds in basic solution

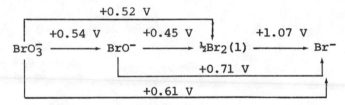

to answer the following:
(a) Does $BrO^-$ disproportionate in basic solution?
(b) Does $Br_2$ disproportionate in basic solution?

13. The potential difference between two hydrogen cells ($H_2 \rightarrow 2H^+ + 2e^-$) is 0.076 V. One of the cells contains $1.0M$ $H^+$ and $H_2(g)$ at 1 atm. The other contains 1 atm $H_2(g)$. What is the concentration of $H^+$ in the second cell?

$Pt|H_2|H^+(?M)||H^+(1.0M)|H_2|Pt$

14. Calculate the concentration of $Fe^{2+}$ and $Fe^{3+}$ in a half cell that contains a total iron concentration of $1.00 \times 10^{-3}M$ if the potential difference of the cell with respect to a standard hydrogen electrode is -0.694 V.

$Pt|Fe^{2+}(?M)||H^+(1M)|H_2(1\text{ atm})|Pt$

15. For 3.00 hours a current of 10.0 amp plated out 13.6 g of a substance. What is the equivalent weight of the substance?

16. In a fuel cell hydrogen and oxygen can react to produce electricity. The process is represented by the cell in Figure 10.1 of the study guide.

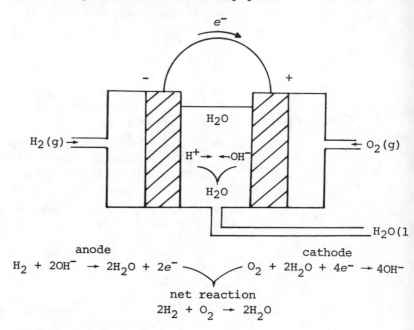

anode

$$H_2 + 2OH^- \rightarrow 2H_2O + 2e^-$$

cathode

$$O_2 + 2H_2O + 4e^- \rightarrow 4OH^-$$

net reaction

$$2H_2 + O_2 \rightarrow 2H_2O$$

FIGURE 10.1   Fuel cell.

In the process hydrogen gas is oxidized to water at the anode, and oxygen gas is reduced to $OH^-$ at the cathode. The excess water is removed. Answer the following:

(a) If 47.8 liters of hydrogen at STP react in 10.0 minutes, how many liters of oxygen at STP react in the same time?

(b) How many moles of water are produced in 10.0 minutes?

(c) What is the average current produced during this time?

17. Wind generators are attractive sources of energy except for the fact that the wind occasionally ceases to blow. It has been proposed that the generator be used to electrolyze water during off-peak power times during these hours and that the hydrogen and oxygen

generated during these hours could be used in a fuel cell to generate electricity when the wind stops blowing. If the efficiency is 100 percent—i.e., no energy losses occur—what size of tank would be needed to store sufficient hydrogen and oxygen at STP to supply a single residence with 20 amps for 5.0 hours?

III. Do the following for each of the electrochemical cells shown on pages 166, 167, and 168.

(a) Label the anode and the cathode.
(b) Write the reaction occurring at the anode and that occurring at the cathode.
(c) Show the direction of the electron movement in the external connection.
(d) Indicate the sign of the cathode and that of the anode.
(e) Show the direction of ion movement in the cell.
(f) Diagram the cell in the shorthand notation described in Section 10.7 of your text. Assume that all concentrations are $1M$.
(g) Determine the half-cell reactions from the standard electrode potentials in Table 10.1 of the study guide.
(h) Calculate the cell potential.

*Note*: Electrochemical cells 1, 2, and 3 are on pages 166, 167, and 168 of the study guide.

## ANSWERS TO EXERCISES

I. Basic electrochemical principles

1. True [10.2]

2. False [10.2]    In a galvanic cell the positive electrode is the cathode, and reduction occurs at this electrode. The conventions relating to the terms anode and cathode are outlined in Table 10.2 of the study guide.

TABLE 10.2  Electrode conventions

|  | Cathode | Anode |
|---|---|---|
| ions attracted | cations | anions |
| direction of electron movement | into cell | out of cell |
| half reaction | reduction | oxidation |
| sign |  |  |
| electrolysis cell | negative | positive |
| galvanic cell | positive | negative |

1.

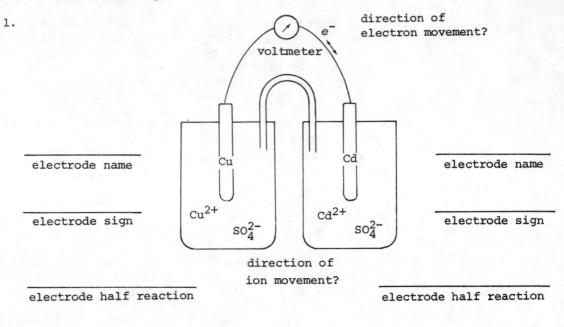

direction of
electron movement?

voltmeter

electrode name _____

electrode sign _____

electrode half reaction _____

standard electrode
potential _____

direction of
ion movement?

electrode name _____

electrode sign _____

electrode half reaction _____

standard electrode
potential _____

cell notation _____

overall cell reaction _____

cell potential _____

2.

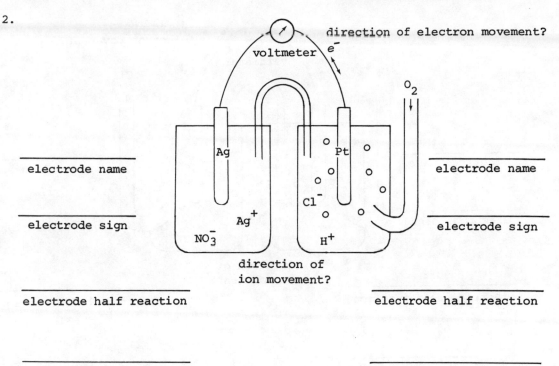

direction of electron movement?

voltmeter

$O_2$

Ag

Pt

$Cl^-$

$Ag^+$

$H^+$

$NO_3^-$

electrode name _____                    electrode name _____

electrode sign _____                     electrode sign _____

direction of
ion movement?

electrode half reaction _____          electrode half reaction _____

standard electrode
potential _____                        standard electrode
                                                 potential _____

cell notation _____

overall cell reaction _____

cell potential _____

3.

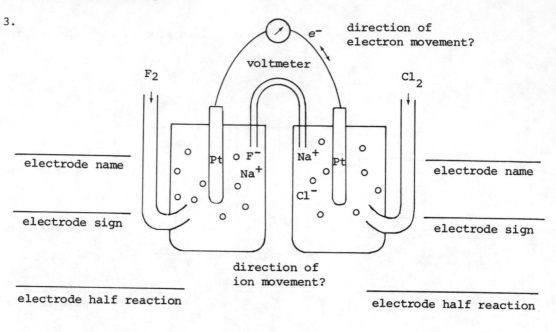

direction of
electron movement?

voltmeter

$F_2$                                $Cl_2$

electrode name _____                    _____ electrode name

electrode sign _____                    _____ electrode sign

_____                              direction of              _____
electrode half reaction                    ion movement?            electrode half reaction

_____                                                        _____
standard electrode potential                                        standard electrode potential

_____
cell notation

_____
overall cell reaction

_____
cell potential

You may find it easier to remember the conventions relating to a galvanic, or voltaic, cell by using the mnemonic device outlined in Table 10.3 of the study guide. Note that when the opposite electrochemical terms are alphabetized, the first term describes the anode and the second describes the cathode.

For an electrolysis cell, i.e., a cell in which an external power supply causes the reaction to proceed in a nonspontaneous direction, the mnemonic device is valid except that the electrode signs are reversed. In an electrolysis cell, oxidation occurs at the positive electrode, and electrons move away from this electrode in the external circuit. Also, cell notations are not used for electrolysis cells.

### TABLE 10.3  Electrode Conventions for Galvanic Cells

|  | Anode | Cathode |
|---|---|---|
| ions attracted | anions | cations |
| electrode sign | negative | positive |
| electrode process | oxidation | reduction |
| direction of electron movement | away | toward |
| location in cell notation | left | right |

3. True [10.8]

4. True [10.2]

5. True [10.8]

6. False [10.12]   Batteries are sources of energy and are therefore voltaic cells. Electrolytic cells require an energy source to operate.

7. False [10.4]   The balanced equation

$$Cu^{2+} + 2e^- \rightarrow Cu(s)$$

indicates that 2 mol of electrons, or 2 F, are required to produce 1 mol of Cu(s).

8. True [10.4]

9. True [10.8]   The oxygen-producing half reaction

$$2H_2O \rightarrow O_2 + 4H^+ + 4e^-$$

indicates that 4 faradays are required to produce 1 mol of $O_2$, or 22.4 liters of $O_2$, at STP.

10. True [10.9]   Notice that the Nernst equation

$$\mathcal{E} = \mathcal{E}^\circ - \frac{RT}{nF} \ln Q$$

includes temperature, $T$. The standard electrode potentials in Table 10.1 of the study guide are given for a specific temperature, 25°C.

11. True [10.8]

12. False [10.8]  A positive cell emf indicates that the reaction occurs spontaneously as written. In Table 10.1 of the study guide, oxidizing agents are on the left. The stronger the oxidizing agent, the more positive the value of $\mathscr{E}°$. Reducing agents are on the right. The stronger the reducing agent, the more negative the value of $\mathscr{E}°$. Thus, we diagram Table 10.1 of the study guide as follows:

increasing
strength
of
oxidizing
agent

| oxidizing $\rightarrow$ reducing |
| agent        agent |

increasing
strength
of
reducing
agent

The best reducing agent in Table 10.1 of the study guide in Li. The $Li^+$ ion is reduced; thus, it is an oxidizing agent with an $\mathscr{E}°$ of -3.045 V. The lithium atom is a reducing agent, being oxidized to $Li^+$ with an $\mathscr{E}°$ of +3.045 V.

13. True [10.1]  Ohm's law states

$$\mathscr{E} = IR$$

The proportionality constant $R$ is called the resistance.

14. True [10.3]

15. False [10.4]  According to the reaction

$$2H^+ + 2e^- \rightarrow H_2$$

1 F, i.e., 1 mol of electrons, would liberate $\frac{1}{2}$ mol of $H_2$, or $\frac{1}{2}$ mol $H_2$ (2 g $H_2$/1 mol $H_2$) = 1 g $H_2$.

16. True [10.8]

17. False [10.8]  The cell emf is the sum of the half-cell potential for the oxidation reaction and the half-cell potential for the reduction reaction. In the table of standard electrode potentials (Table 10.1 of the study guide), all reactions are written as reduction reactions. Thus, both the sign of one electrode potential and the corresponding reaction must be reversed to obtain the necessary half-cell potention and oxidation reaction.

18. False [10.8]    By convention, the anode is indicated on the left. The cell reaction as diagrammed is

$$Cd(s) + Ni^{2+} \rightarrow Cd^{2+} + Ni(s)$$

Thus, Ni is deposited on the electrode, the Ni electrode gains weight, and the Cd electrode loses weight.

19. True [10.9]    The balanced equation

$$2Cr + 3Fe^{2+} \rightarrow 2Cr^{3+} + 3Fe$$

indicated that six electrons are exchanged, $n = 6$, and that

$$Q = \frac{[Cr^{3+}]^2}{[Fe^{2+}]^3} = \frac{[10^{-3}]^2}{[10^{-2}]^3}$$

The metals Cr and Fe have unit activity and are not included in $Q$. The value of $\mathscr{E}°$ is +0.304 V:

$$
\begin{array}{ll}
3[Fe^{2+} + 2e^- \rightleftharpoons Fe] & \mathscr{E}° = -0.4402 \text{ V} \\
2[Cr \rightleftharpoons Cr^{3+} + 3e^-] & \mathscr{E}° = +0.744 \text{ V} \\
\hline
2Cr + 3Fe^{2+} \rightarrow 2Cr^{3+} + 3Fe & \mathscr{E}° = +0.304 \text{ V}
\end{array}
$$

20. False [10.11]    Higher voltage must be applied if concentration polarization occurs. As electrolysis proceeds, there is a buildup or deficit of ions around the electrode, thus setting up a concentration cell about the electrode. The concentration cell produces a back emf, opposite to the applied voltage.

21. True [10.1]

22. False [10.4]    One faraday equals 96,487 coulombs.

## II. Electrochemical calculations

1. +0.293 V [10.7]    First write the necessary half reactions, balance them, and add them. Multiply any equation and the corresponding value of $\Delta G$ by integers so that unwanted intermediate ions or molecules cancel. The desired result is thus obtained in the same manner as you used in the law of Hess.

$$Cr_2O_7^{2-} + 14H^+ + 6e^- \rightleftharpoons 2Cr^{3+} + 7H_2O$$
$$\Delta G = -6F(+1.33 \text{ V})$$

$$2Cr^{3+} + 6e^- \rightleftharpoons 2Cr$$
$$\Delta G = 2(-3F(0.744 \text{ V}))$$
$$= -4.464F$$

$$\overline{Cr_2O_7^{2-} + 14H^+ + 12e^- \rightleftharpoons 2Cr + 7H_2O}$$
$$\Delta G = -6F(0.58 \text{ V})$$
$$= -3.516F$$

Rearranging so that 12 electrons appear in the equation for $\Delta G$:

$$\Delta G = -nF\mathscr{E}°$$
$$-3.516F = -12F\mathscr{E}°$$
$$\mathscr{E}° = +0.293 \text{ V}$$

**2. rust**
**[10.8]**

For the tin-coated iron, possible oxidation reactions (see Table 10.1 of the study guide), are

$$Fe \rightarrow Fe^{2+} + 2e^- \qquad \mathscr{E}° = +0.4402 \text{ V}$$
$$Sn \rightarrow Sn^{2+} + 2e^- \qquad \mathscr{E}° = +0.136 \text{ V}$$
$$2H_2O \rightarrow O_2 + 4H^+ + 4e^- \qquad \mathscr{E}° = -1.229 \text{ V}$$

The reduction reaction is

$$O_2 + 2H_2O + 4e^- \rightarrow 4OH^- \qquad \mathscr{E}° = +0.401 \text{ V}$$

Since iron is the best of the three reducing agents, it will be oxidized—i.e., it will rust—when not protected.

**3. no rust**
**[10.8]**

For galvanized iron, possible oxidation reactions (see Table 10.1 of the study guide), are

$$Fe \rightarrow Fe^{2+} + 2e^- \qquad \mathscr{E}° = +0.4402 \text{ V}$$
$$Zn \rightarrow Zn^{2+} + 2e^- \qquad \mathscr{E}° = +0.7628 \text{ V}$$
$$2H_2O \rightarrow O_2 + 4H^+ + 4e^- \qquad \mathscr{E}° = -1.229 \text{ V}$$

The reduction reaction is

$$O_2 + 2H_2O + 4e^- \rightarrow 4OH^- \qquad \mathscr{E}° = +0.401 \text{ V}$$

The iron is protected since zinc is a better reducing agent than iron. The zinc will be oxidized by the $O_2$ and the $H_2O$.

**4. 2**
**[10.7]**

We use the relationship between standard Gibbs free energy and standard emf:

$$\Delta G° = -nF\mathscr{E}°$$

$$n = \frac{-\Delta G°}{F\mathscr{E}°}$$

$$n = \frac{-(-50.21 \text{ kJ/mol})}{(96.487 \text{ kJ/V})(+0.26 \text{ V})}$$

$$n = 2.0$$

**5. (a) 1.1 hr**
**[10.4]**

The reduction reaction is

$$Mg^{2+} + 2e^- \rightarrow Mg$$

Thus, 2 mol of electrons, or 2 F, are required per mole of Mg. We can use a logic chain to solve the problem:

? hr

$$= 5.0 \text{ kg Mg} \left(\frac{10^3 g}{1 \text{ kg}}\right)\left(\frac{1 \text{ mol Mg}}{24.3 \text{ g Mg}}\right)\left(\frac{2F}{1 \text{ mol Mg}}\right)\left(\frac{96,500 \text{ coulombs}}{1F}\right)$$

$$\times \left(\frac{1}{1.0 \times 10^4 \text{ coulomb/sec}}\right)\left(\frac{1 \text{ min}}{60 \text{ sec}}\right)\left(\frac{1 \text{ hr}}{60 \text{ min}}\right)$$

$$= 1.1 \text{ hr}$$

(b) $4.6 \times 10^3$ liters $Cl_2$

The oxidation reaction is

$$2Cl^- \rightarrow Cl_2 + 2e^-$$

Thus, for every mole of Mg produced, 1 mol of $Cl_2$, or 22.4 liters of $Cl_2$, is obtained. Therefore,

? liters $Cl_2$

$$= 5.0 \text{ kg Mg}\left(\frac{10^3 g}{1 \text{ kg}}\right)\left(\frac{1 \text{ mol Mg}}{24.3 \text{ g Mg}}\right)\left(\frac{1 \text{ mol Cl}_2}{1 \text{ mol Mg}}\right)\left(\frac{22.4 \text{ liters Cl}_2}{1 \text{ mol Cl}_2}\right)$$

$$= 4.6 \times 10^3 \text{ liters Cl}_2$$

We can also solve the problem as follows:

? liters $Cl_2$

$$= 1.1 \text{ hr}\left(\frac{60 \text{ min}}{1 \text{ hr}}\right)\left(\frac{60 \text{ sec}}{1 \text{ min}}\right)\left(\frac{1.0 \times 10^4 \text{ coulombs}}{1 \text{ sec}}\right)$$

$$\times \left(\frac{1F}{96,500 \text{ coulombs}}\right)\left(\frac{1 \text{ mol Cl}_2}{2F}\right)\left(\frac{22.4 \text{ liters Cl}_2}{1 \text{ mol Cl}_2}\right)$$

$$= 4.6 \times 10^3 \text{ liters Cl}_2$$

6. [10.8]    Spontaneous reactions are reactions with a positive cell potential.

(a) no    Reaction (a) is not spontaneous since $\mathcal{E}^\circ_{cell} = -0.153$ V:

| | |
|---|---|
| $Cd^{2+} + 2e^- \rightarrow Cd$ | $\mathcal{E}^\circ = -0.4029$ V |
| $Ni \rightarrow Ni^{2+} + 2e^-$ | $\mathcal{E}^\circ = +0.250$ V |
| $Cd^{2+} + Ni \rightarrow Cd + Ni^{2+}$ | $\mathcal{E}^\circ_{cell} = -0.153$ V |

(b) yes    Reaction (b) is spontaneous since $\mathcal{E}^\circ_{cell} = +0.473$ V:

| | |
|---|---|
| $Sn \rightarrow Sn^{2+} + 2e^-$ | $\mathcal{E}^\circ = +0.136$ V |
| $Cu^{2+} + 2e^- \rightarrow Cu$ | $\mathcal{E}^\circ = +0.337$ V |
| $Sn + Cu^{2+} \rightarrow Sn^{2+} + Cu$ | $\mathcal{E}^\circ_{cell} = +0.473$ V |

(c) no    Reaction (c) is not spontaneous since $\mathcal{E}^\circ_{cell} = -0.2636$ V:

| | |
|---|---|
| $I_2 + 2e^- \rightarrow 2I^-$ | $\mathcal{E}^\circ = +0.5355$ V |
| $2Ag \rightarrow 2Ag^+ + 2e^-$ | $\mathcal{E}^\circ = -0.7991$ V |
| $I_2 + 2Ag \rightarrow 2I^- + 2Ag^+$ | $\mathcal{E}^\circ_{cell} = -0.2636$ V |

Note that the half reaction for the oxidation of Ag must be multiplied by 2 before addition so that the electrons lost and gained in the half reactions will cancel. The $\mathcal{E}°$ for the Ag/Ag$^+$ electrode, however, is *not* multiplied by 2. The $\mathcal{E}°$ value of a half reaction is independent of the number of electrons lost or gained.

(d) yes

Reaction (d) is spontaneous since

$$2[MnO_4^- + 8H^+ + 5e^- \rightarrow Mn^{2+} + 4H_2O] \qquad \mathcal{E}° = +1.51 \text{ V}$$
$$\underline{5[2Br^- \rightarrow Br_2 + 2e^-] \qquad\qquad\qquad\qquad \mathcal{E}° = -1.0562 \text{ V}}$$
$$2MnO_4^- + 16H^+ + 10Br^- \rightarrow 2Mn^{2+} + 8H_2O + 5Br_2 \quad \mathcal{E}°_{cell} = +0.44 \text{ V}$$

7. [10.8]

In standard cell notation the anode is listed first. Oxidation always occurs at the anode. Thus, for reaction (b) the cell is noted

(b) $Sn|Sn^{2+}||Cu^{2+}|Cu$

and for reaction (d) the cell is noted

(d) $Pt|Br^-|Br_2||MnO_4^-|Mn^{2+}|Pt$

The inert platinum electrodes are noted in the cell notation for reaction (d) since they are necessary to make an external electrical connection.

8. +0.079 V
   [10.9]

From the Nernst equation

$$\mathcal{E} = \mathcal{E}° - \frac{0.0592}{n} \log$$

$$\mathcal{E} = 0.00 - 0.0592 \log \frac{(0.010M)(0.071M)}{(0.13M)(0.12M)}$$

$$= -0.0592 \log (4.55 \times 10^{-2})$$

$$= -0.0592(-1.342)$$

$$= +0.079 \text{ V}$$

9. [10.8]

From Table 10.1 of the study guide we choose reactions that contain any of the possible reactants, Cd, Fe$^{3+}$, and H$_2$O:

### Oxidation Reactions

$$Cd \rightarrow Cd^{2+} + 2e^- \qquad\qquad \mathcal{E}° = +0.4029 \text{ V}$$

$$2H_2O \rightarrow O_2 + 4H^+ + 4e^- \qquad \mathcal{E}° = -1.229 \text{ V}$$

### Reduction Reactions

$$Fe^{3+} + e^- \rightarrow Fe^{2+} \qquad\qquad \mathcal{E}° = +0.771 \text{ V}$$

$$Fe^{2+} + 2e^- \rightarrow Fe \qquad\qquad \mathcal{E}° = -0.4402 \text{ V}$$

$$2H_2O + 2e^- \rightarrow H_2 + 2OH^- \qquad \mathcal{E}° = 0.82806 \text{ V}$$

The only pair of reactions that has a positive cell potential is

$$Cd \rightarrow Cd^{2+} + 2e^- \qquad \mathcal{E}° = +0.4029 \text{ V}$$
$$\underline{2[Fe^{3+} + e^- \rightarrow Fe^{2+}] \qquad \mathcal{E}° = +0.771 \text{ V}}$$
$$Cd + 2Fe^{3+} \rightarrow Cd^{2+} + 2Fe^{2+} \qquad \mathcal{E}°_{cell} = +1.174 \text{ V}$$

10. [10.8]

In Appendix D of your text the reduction potential of $Co^{3+}$ is given:

$$Co^{3+} + e^- \rightarrow Co^{2+} \qquad \mathcal{E}° = +1.808 \text{ V}$$

The oxidation potential of water can be obtained from Table 10.1 of the study guide:

$$2H_2O \rightarrow O_2 + 4H^+ + 4e^- \qquad \mathcal{E}° = -1.229 \text{ V}$$

Thus, when we couple the reduction of $Co^{3+}$ and the oxidation of water, we find that $\mathcal{E}°$ for the oxidation—reduction reaction is positive:

$$4[Co^{3+} + e^- \rightarrow Co^{2+}] \qquad \mathcal{E}° = +1.808 \text{ V}$$
$$\underline{2H_2O \rightarrow O_2 + 4H^+ + 4e^- \qquad \mathcal{E}° = -1.229 \text{ V}}$$
$$4Co^{3+} + 2H_2O \rightarrow 4Co^{2+} + O_2 + 4H^+ \qquad \mathcal{E}° = +0.579 \text{ V}$$

Thus, when the $Co^{3+}$ salt is added to water, the reduction of $Co^{3+}$ and the oxidation of water occur spontaneously.

11. 99.8 g Cu
[10.4]

The balanced equation is

$$2Cu^{2+} + 2H_2O \rightarrow 2Cu + O_2 + 4H^+$$

Using a logic chain, we find

$$? \text{ g Cu} = 17.6 \text{ liters } O_2 \left(\frac{1 \text{ mol } O_2}{22.41 \text{ liters } O_2}\right)\left(\frac{2 \text{ mol Cu}}{1 \text{ mol } O_2}\right)\left(\frac{63.55 \text{ g Cu}}{1 \text{ mol Cu}}\right)$$

$$= 99.8 \text{ g Cu}$$

12. [10.8]

(a) yes

$$2[BrO^- + 2e^- + 2H^+ \rightarrow Br^- + H_2O] \qquad \mathcal{E}° = +0.71 \text{ V}$$
$$\underline{BrO^- + 2H_2O \rightarrow BrO_3^- + 4H^+ + 4e^- \qquad \mathcal{E}° = -0.54 \text{ V}}$$
$$3BrO^- \rightarrow 2Br^- + BrO_3^- \qquad \mathcal{E}° = +0.17 \text{ V}$$

(b) yes

$$5[\tfrac{1}{2}Br_2 + e^- \rightarrow Br^-] \qquad \mathcal{E}° = +1.07 \text{ V}$$
$$\underline{\tfrac{1}{2}Br_2 + 6OH^- \rightarrow BrO_3^- + 3H_2O + 5e^- \qquad \mathcal{E}° = -0.52 \text{ V}}$$
$$3Br_2 + 6OH^- \rightarrow 5Br^- + BrO_3^- + 3H_2O \qquad \mathcal{E}° = +0.55 \text{ V}$$

13. $5.2 \times 10^{-2} M$
[10.9]

The reaction taking place

$$H_2 (1 \text{ atm}) + 2H^+(1.0M) \rightarrow H_2 (1 \text{ atm}) + 2H^+(?M) \quad \mathcal{E} = +0.076 \text{ V}$$

Two faradays are involved in the cell reaction, and therefore $n \doteq 2$. Since the same electrode is in each half cell, $\mathcal{E}°$ for the cell is zero.

We can obtain the unknown $[H^+]$ from the Nernst equation:

$$\mathcal{E} = \mathcal{E}° - \frac{0.0592}{n} \log Q$$

$$0.076 = 0 - \frac{0.0592}{2} \log \left( \frac{(1) \, [H^+]^2}{(1) \, (1)^2} \right)$$

$$\log \, [H^+]^2 = 0.076 \left( \frac{2}{0.0592} \right)$$

$$= -2.57$$

$$[H^+]^2 = \text{antilog} \; -2.57$$

$$= 2.7 \times 10^{-3}$$

$$[H^+] = 5.2 \times 10^{-2} M$$

14. [10.9]
$[Fe^{2+}]$
$= 9.52 \times 10^{-4} M$
$[Fe^{3+}]$
$= 4.8 \times 10^{-5} M$

Since the total iron concentration is $1.00 \times 10^{-3}$ ,

$$[Fe^{2+}] + [Fe^{3+}] = 1.00 \times 10^{-3}$$

If we let $x = [Fe^{2+}]$, then $((1.00 \times 10^{-3}) - x)M = [Fe^{3+}]$.
We can now use the Nernst equation to solve the problem:

$$\mathcal{E} = \mathcal{E}° - \frac{0.0592}{n} \log Q$$

$$\log Q = (\mathcal{E} - \mathcal{E}°) \, \frac{-n}{0.0592}$$

$$\log \frac{((1.00 \times 10^{-3}) - x)M}{x} = (-0.694 - (0.771)) \, \frac{-1}{0.0592}$$

$$x = 9.52 \times 10^{-4} M = [Fe^{2+}]$$

$$((1.00 \times 10^{-3}) - x)M = 4.8 \times 10^{-5} M = [Fe^{3+}]$$

15. 12.1 g/equiva-
lent
[10.4]

We can use Faraday's laws and a logic chain to solve this
problem:

$$? \text{ equivalents} = 3.00 \text{ hr} \left( \frac{60 \text{ min}}{1 \text{ hr}} \right) \left( \frac{60 \text{ sec}}{1 \text{ min}} \right)$$

$$\times \, (10.0 \text{ coulomb/sec}) \left( \frac{1 \text{ F}}{96.487 \text{ coulombs}} \right) \left( \frac{1 \text{ equivalent}}{1 \text{ F}} \right)$$

$$= 1.12 \text{ equivalents}$$

$$\text{equivalent weight} = \frac{\text{weight}}{\text{number of equivalents}}$$

$$= \frac{13.6 \text{ g}}{1.12 \text{ equivalents}}$$

$$= 12.1 \text{ g/equivalent}$$

16. (a) 23.9 li-
    ters $O_2$
    [10.4]

The net reaction shows that 1 mol of $O_2$ reacts with 2 mol of $H_2$; therefore, half as much $O_2$ as $H_2$, or 23.9 liters of $O_2$, is required at STP:

? liters $O_2$

$$= 47.8 \text{ liters } H_2\left(\frac{1 \text{ mol } O_2}{2 \text{ mol } H_2}\right)\left(\frac{22.4 \text{ liters } O_2/1 \text{ mol } O_2}{22.4 \text{ liters } H_2/1 \text{ mol } H_2}\right)$$

$$= 23.9 \text{ liters } O_2$$

(b) 2.13 mol
    $H_2O$

From the net reaction we know that 2 mol of $H_2$ reacts with 1 mol of $O_2$ to produce 2 mol of $H_2O$. Thus,

? liters $H_2O$

$$= 47.8 \text{ liters } H_2\left(\frac{2 \text{ mol } H_2O}{2 \text{ mol } H_2}\right)\left(\frac{1 \text{ mol } H_2}{22.4 \text{ liters } H_2}\right)$$

$$= 2.13 \text{ mol } H_2O$$

(c) 686 cou-
    lombs/sec

To determine the current, expressed in amperes or coulombs/sec, we use Faraday's laws and a logic chain:

? coulomb/sec

$$= \left(\frac{47.8 \text{ liters } H_2}{10 \text{ min}}\right)\left(\frac{1 \text{ mol } H_2}{22.4 \text{ liters } H_2}\right)\left(\frac{2 \text{ F}}{1 \text{ mol } H_2}\right)$$

mol $H_2$
reacted
per min

$$\times \left(\frac{96,500 \text{ coulombs}}{1 \text{ F}}\right)\left(\frac{1 \text{ min}}{60 \text{ sec}}\right)$$

current in
coulombs/sec

$$= 686 \text{ coulombs/sec}$$

17. 42 liters $H_2$
    21 liters $O_2$
    [10.4]

To determine the size of the tanks, we use Faraday's laws and set up a logic chain:

? liters $H_2$

$$= 20 \text{ coulombs/sec}\left(\frac{60 \text{ sec}}{1 \text{ min}}\right)\left(\frac{60 \text{ min}}{1 \text{ hr}}\right)\left(\frac{22.4 \text{ liters } H_2}{1 \text{ mol } H_2}\right)$$

$$\times \left(\frac{1 \text{ mol } H_2}{2 \text{ F}}\right)\left(\frac{1 \text{ F}}{96,500 \text{ coulombs}}\right)(5 \text{ hr})$$

$$= 42 \text{ liters } H_2$$

Since the cell reaction

$$2H_2(g) + O_2(g) \rightarrow 2H_2O(l)$$

indicates that 1 mol of $O_2$ reacts with 2 mol of $H_2$, only 21 liters of $O_2$ are required.

III. Electrochemical cells

1.

In solving problems such as this, check the table of standard electrode potential (table 10.1 of the study guide) first to determine which species will be oxidized and which will be reduced. The couple with the more positive electrode potential will occur as written, a reduction; the other will occur in reverse, an oxidation.

In this cell $Cu^{2+}$ is reduced and Cd is oxidized. The electrons travel toward the positive electrode. The ions that are reduced, the cations, migrate toward the cathode, and the anions migrate toward the anode. In the cell notation the anode is represented to the left of the salt bridge, which is represented by the double bar, and the cathode is represented to the right of the salt bridge. The cell potential is

$$\mathcal{E}^{\circ}_{right} - \mathcal{E}^{\circ}_{left} = +0.740 \text{ V}$$

Note: Refer to the diagram on page 179.

1.

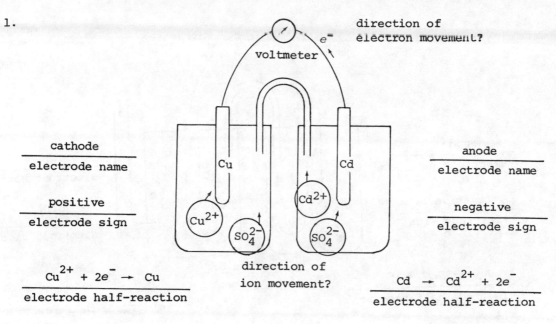

direction of
electron movement?

voltmeter

$e^-$

cathode
―――――――
electrode name

positive
―――――――
electrode sign

$Cu^{2+} + 2e^- \rightarrow Cu$
――――――――――――――――
electrode half-reaction

+0.337$V$
―――――――
standard electrode
potential

anode
―――――――
electrode name

negative
―――――――
electrode sign

$Cd \rightarrow Cd^{2+} + 2e^-$
――――――――――――――――
electrode half-reaction

+0.4029$V$
―――――――
standard electrode
potential

direction of
ion movement?

$Cd \mid Cd^{2+} \parallel Cu^{2+} \mid Cu$
―――――――――――――――――
cell notation

$Cd + Cu^{2+} \rightarrow Cd^{2+} + Cu$
―――――――――――――――――
overall cell reaction

+0.740$V$
―――――――
cell potential

2.

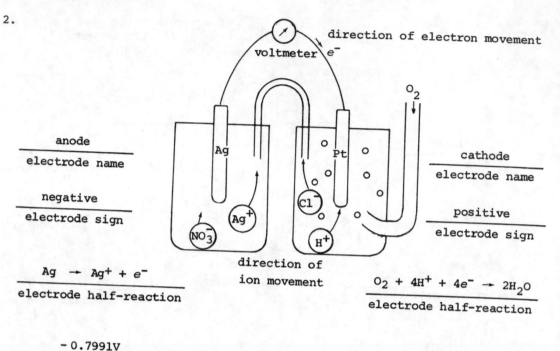

anode
_____
electrode name

negative
_____
electrode sign

$Ag \rightarrow Ag^+ + e^-$
_____
electrode half-reaction

$-0.7991V$
_____
standard electrode
potential

cathode
_____
electrode name

positive
_____
electrode sign

$O_2 + 4H^+ + 4e^- \rightarrow 2H_2O$
_____
electrode half-reaction

$+1.229V$
_____
standard electrode
potential

$Ag \mid Ag^+ \parallel H^+ \mid O_2 \mid Pt$
_____
cell notation

$4Ag + O_2 + 4H^+ \rightarrow 4Ag^+ + 2H_2O$
_____
overall cell reaction

$+0.430V$
_____
cell potential

3.

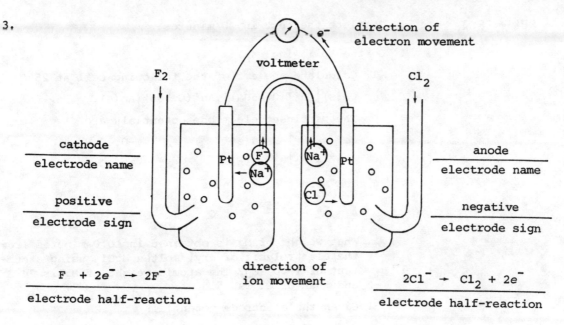

direction of
electron movement

voltmeter

$F_2$

$Cl_2$

| cathode | anode |
|---|---|
| electrode name | electrode name |

| positive | negative |
|---|---|
| electrode sign | electrode sign |

direction of
ion movement

| $F + 2e^- \rightarrow 2F^-$ | $2Cl^- \rightarrow Cl_2 + 2e^-$ |
|---|---|
| electrode half-reaction | electrode half-reaction |

| +2.87 | -1.3595 |
|---|---|
| standard electrode potential | standard electrode potential |

$$Pt \mid Cl_2 \mid Cl^- \parallel F^- \mid F_2 \mid Pt$$
cell notation

$$F_2 + 2Cl^- \rightarrow Cl_2 + 2F^-$$
overall cell reaction

+1.51V
cell potential

SELF-TEST        Complete the test in 20 minutes:

I. Do the following:

1. Calculate the emf of the following cell at 25°C:

   $Cd(s) | Cd^{2+}(0.050M) || Ag^+(0.50M) | Ag(s)$

   The pertinent electrode potentials are

   $2e^- + Cd^{2+} \rightleftharpoons Cd(s)$         $\mathcal{E}° = -0.40$ V

   $e^- + Ag^+ \rightleftharpoons Ag(s)$         $\mathcal{E}° = +0.80$ V

   The Nernst equation is

   $$\mathcal{E} = \mathcal{E}° - \frac{0.0592}{n} \log Q$$

2. What weight of Mg is obtained in 10.0 minutes from the electrolysis of dry, molten $MgCl_2$ using a current of 10.0 amp? The atomic weight of Mg is 24.3 and that of Cl is 35.5.

3. Given the electrode potentials

   $Mg^{2+} + 2e^- \rightleftharpoons Mg$         $\mathcal{E}° = -2.36$ V

   $Cu^{2+} + 2e^- \rightleftharpoons Cu$         $\mathcal{E}° = +0.34$ V

   answer each of the following:
   (a) What is the best oxidizing agent of the chemical species in the half reactions?
   (b) In a voltaic cell consisting of standard $Mg^{2+}/Mg$ and $Cu^{2+}/Cu$ half cells, which half cell is the cathode?
   (c) Is the Cu electrode of the voltaic cell in part (b) above positive or negative?
   (d) Will Mg metal spontaneously reduce $Cu^{2+}$ ions in 1.0M concentration at 25°C?
   (e) In an electrolytic cell is the cathode positive or negative?

II. Answer each of the following. The questions refer to a galvanic cell that involves the reaction

   $$2VO^{2+} + 4H^+ + Ni(s) \rightarrow 2V^{3+} + Ni^{2+} + 2H_2O$$

   Pertinent electrode potentials are

   $2e^- + Ni^{2+} \rightleftharpoons Ni$         $\mathcal{E}° = -0.25$ V

   $e^- + 2H^+ + VO^{2+} \rightleftharpoons V^{3+} + H_2O$         $\mathcal{E}° = +0.36$ V

1. The standard emf of the cell is
   (a) 0.11 V                        (c) 0.97 V
   (b) 0.61 V                        (d) 0.47 V

2. The emf of the cell could be increased by
   (a) increasing the $Ni^{2+}$ concentration
   (b) lowering the $pH$
   (c) reducing $VO^{2+}$ concentration
   (d) increasing $V^{3+}$ concentration

3. The anode could be made of
   (a) Ni                           (c) Pt
   (b) V                            (d) $H_2$

4. The cathode could be made of
   (a) Ni                           (c) Pt
   (b) V                            (d) $H_2$

5. The cell must be constructed so that the mixing of
   the following is avoided:
   (a) $V^{3+}$ and $Ni^{2+}$
   (b) $V^{3+}$ and Ni(s)
   (c) $VO^{2+}$ and Ni(s)
   (d) $VO^{2+}$ and Ni(s)

6. The best oxidizing agent(s) is (are)
   (a) $Ni^{2+}$                    (c) $VO^{2+}$ and $H^+$
   (b) Ni(s)                        (d) $V^{3+}$

# The Nonmetals, Part I

OBJECTIVES

    I. You should be familiar with the physical properties of the group 0 elements (the noble gases), the groups VII A elements (the halogens), and the group VI A elements (except oxygen, see Chapter 8 of your text). The elements discussed in Chapter 11 of your text are shown in the shaded area of Figure 11.1 of the study guide.

    II. You should be familiar with the abundance of these elements in nature, the ways in which they can be obtained from natural sources, and their uses.

    III. You should know the fundamental chemistry of these elements, i.e., the types of compounds they form, how such compounds are prepared, and how such compounds react.

    IV. You should be able to apply the basic information of all previous chapters of your text toward an understanding of the chemistry of these elements.

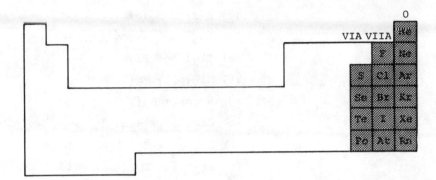

FIGURE 11.1  Periodic table.

**EXERCISES**

I. Answer the following:

1. Chlorine can be prepared by the reaction of concentrated HCl with solid $MnO_2$. Write the balanced equation for the reaction:

   $MnO_2(s) + HCl(aq) \rightarrow$

2. Use the ion-electron method to balance the equation for the reaction of iodate ion and sulfite ion to form iodine and sulfate ion in acid solution.

3. What is the physical state at STP of each of the following: fluorine, chlorine, bromine, and iodine?

4. Write the electronic configuration of each of the following:
   (a) Cl          (b) $Cl^-$          (c) Br          (d) $Br^-$

5. From the following standard electrode potential diagram of chlorine, chloride, and the chlorine oxyanions, predict which compounds are stable in basic solution:

$$\overset{\text{+.63 V}}{\overbrace{ClO_4^- \xrightarrow{\text{+.36 V}} ClO_3^- \xrightarrow{\text{+.33 V}} ClO_2^- \xrightarrow{\text{+.66 V}} ClO^- \xrightarrow{\text{+.40 V}} Cl_2 \xrightarrow{\text{+1.36 V}} Cl^-}}$$

$$\underset{\text{+.50 V}}{\underbrace{ClO_3^- \cdots ClO^-}} \qquad \underset{\text{+.89 V}}{\underbrace{ClO^- \cdots Cl^-}}$$

6. Complete each of the following by writing a balanced equation. If no reaction occurs, write *NR* in the space provided for products:

   (a) $SiO_2(s) + 6HF(aq) \rightarrow$

   (b) $Br_2(l) + 2I^-(aq) \rightarrow$

(c) $S(s) + H_2(g) \rightarrow$

(d) $Ba^{2+}(aq) + SO_4^{2-}(aq) \rightarrow$

(e) $Cl_2(g) + Bi(s) \rightarrow$

(f) $Cl_2(g) + Hg(l) \rightarrow$

(g) $Se(s) + Hg(l) \rightarrow$

(h) $CaF_2(s) + H_2SO_4(g) \xrightarrow{\text{heat}}$

(i) $Br_2(g) + H_2(g) \rightarrow$

(j) $Br_2(l) + 2Cl^-(aq) \rightarrow$

7. What is the orbital hybridization of sulfur in each of the following? Draw structures of the molecules.
   (a) $SF_4$                     (b) $SF_6$

8. Predict the shape of each of the following inter-halogens:
   (a) $ClF_2^+$        (b) $ICl_2^-$        (c) $ClF_3$        (d) $ICl_4^-$

9. How are pure neon, argon, krypton, and xenon obtained commercially?

10. What is the commercial source of pure helium?

11. Why have $XeF_2$, $XeF_4$, and $XeF_6$ been prepared, but not $XeF_3$ and $XeF_5$?

12. In addition to the xenon fluoride compounds mentioned in problem 11 above, some krypton fluoride compounds have been prepared. What are the chances of preparing compounds containing fluorine and either He, Ne, or Ar?

13. Draw the molecular orbital energy-level diagram for diatomic fluorine and explain why $F_2$ is more stable than two free fluorine atoms.

14. Hydrofluoric acid, HF, is a weak electrolyte, but the others of the series HF, HCl, HBr, and HI are strong. Why?

15. What are the color and the physical state of each of the following at STP: oxygen, sulfur, selenium, and tellurium?

16. Write the electronic configuration of tellurium.

17. Draw a Lewis structure of $S_8$.

18. Write the formula for each of the following compounds in the space provided:
    (a) selenous acid
    (b) perbromic acid
    (c) tellurium dioxide

(d)  xenon(VIII) oxide
(e)  calcium fluoride
(f)  sodium bisulfate
(g)  pyrosulfuric acid
(h)  hypoiodous acid
(i)  phosphorus trichloride
(j)  peroxydisulfuric acid
(k)  sodium perxenate
(l)  sodium thiosulfate

19. Determine the oxidation number of iodine in each of
the following compounds. Name each compound in the
space provided.
(a)  HOI
(b)  $IF_7$
(c)  $I_2$
(d)  NaI
(e)  $I_2O_5$
(f)  $H_5IO_6$
(g)  AgI
(h)  $ICl_3$
(i)  $I_2O_4$
(j)  $As_2I_4$
(k)  $I_2Cl_6$

20. Draw a Lewis structure of the thiosulfate ion, $S_2O_3^{2-}$.

21. How many grams of $MnO_2$ are needed to prepare 1.00
liter of dry $Cl_2$ gas measured at 25°C and 1.00 atm
by the following reaction?

$$MnO_2(s) + 4HCl(aq) \rightarrow Cl_2(g) + MnCl_2(aq) + 2H_2O(l)$$

22. What would you expect the shape of the ion $IO_6^{5-}$ to
be? Draw the Lewis structure.

23. Chlorine is contained in DDT, $C_{14}H_9Cl_5$, the common
insecticide. What is the percentage of chlorine in
DDT?

## ANSWERS TO EXERCISES

I. To determine the amount of descriptive chemistry you
should know, rely on your class notes and the comments
of your instructor.

1. [8.5, 11.3]    $MnO_2(s) + 4HCl(aq) \rightarrow Cl_2(g) + MnCl_2(aq) + 2H_2O(l)$

2. [8.5, 11.3]    $2IO_3^- + 2H^+ + 5SO_3^{2-} \rightarrow 5SO_4^{2-} + I_2 + H_2O$

3. [11.2]

| Element | Color | Physical State at STP |
|---|---|---|
| fluorine | pale yellow | gas |
| chlorine | green-yellow | gas |
| bromine | red | liquid |
| iodine | black | solid |

4. 2.15, 2.16,
   11.2

(a) Cl: $1s^22s^22p^63s^23p^5$
(b) Cl⁻: $1s^22s^22p^63s^23p^6$
(c) Br: $1s^22s^22p^63s^23p^63d^{10}4s^24p^5$
(d) Br⁻: $1s^32s^22p^63s^23p^63d^{10}4s^24p^6$

5. 11.7

The chloride ion, $Cl^-$, and the perchlorate ion, $ClO_4^-$, are stable to disproportionation in basic solution. The perchlorate ion is stable because it cannot be further oxidized, and the chloride ion is stable because it cannot be further reduced. Neither of these ions can disproportionate because neither can undergo oxidation and reduction spontaneously and simultaneously. The chlorate ion, $ClO_3^-$, however, can disproportionate:

$$6e^- + 3H_2O + ClO_3^- \rightarrow Cl^- + 6OH^- \qquad \mathcal{E}° = +0.63 \text{ V}$$
$$3[2OH^- + ClO_3^- \rightarrow ClO_4^- + H_2O + 2e] \qquad \mathcal{E}° = -0.36 \text{ V}$$
$$\overline{4ClO_3^- \rightarrow Cl^- + 3ClO_4^-} \qquad \mathcal{E}° = +0.27 \text{ V}$$

Also, the chlorite ion, $ClO_2^-$, can disproportionate:

$$2e^- + ClO_2^- + H_2O \rightarrow ClO^- + 2OH^- \qquad \mathcal{E}° = +0.66 \text{ V}$$
$$ClO_2^- + 2OH^- \rightarrow ClO_3^- + H_2O + 2e^- \qquad \mathcal{E}° = +0.33 \text{ V}$$
$$\overline{2ClO_2^- \rightarrow ClO^- + ClO_3^-} \qquad \mathcal{E}° = +0.33 \text{ V}$$

Both of the products of the disproportionation of $ClO_2^-$ can disproportionate. The $ClO^-$ can disproportionate as follows:

$$2[H_2O + ClO^- + 2e^- \rightarrow Cl^- + 2OH^-] \qquad \mathcal{E}° = +0.89 \text{ V}$$
$$ClO^- + 4OH^- \rightarrow ClO_3^- + 2H_2O + 4e^- \qquad \mathcal{E}° = -0.50 \text{ V}$$
$$\overline{3ClO^- \rightarrow 2Cl^- + ClO_3^-} \qquad \mathcal{E}° = +0.39 \text{ V}$$

Chlorine gas, $Cl_2$, disproportionates as follows:

$$2OH^- + Cl_2 \rightarrow ClO^- + Cl^- + H_2O \qquad \mathcal{E}° = +0.96 \text{ V}$$

6. (a) [11.5]

$$SiO_2(s) + 6HF(aq) \rightarrow 2H^+(aq) + SiF_6^{2-}(aq) + 2H_2O(l)$$

   (b) [11.3]

$$Br_2(l) + 2I^-(aq) \rightarrow 2Br^-(aq) + I_2(s)$$

You should learn some of the important halogen reactions summarized in Table 11.1 of the study guide.

   (c) [11.9]

$$S(s) + H_2(g) \rightarrow H_2S(g)$$

You should learn some of the important sulfur, selenium, and tellurium reactions summarized in Table 11.2 of the study guide.

TABLE 11.1  Some Reactions of the Halogens
(X$_2$ = F$_2$, Cl$_2$, Br$_2$, or I$_2$)

| General Reaction | Remarks |
|---|---|
| $nX_2 + 2M \rightarrow 2MX_n$ | F$_2$, Cl$_2$ with practically all metals; Br$_2$, I$_2$ with all except noble metals |
| $X_2 + H_2 \rightarrow 2HX$ | |
| $3X_2 + 2P \rightarrow 2PX_3$ | With excess P; similar reactions with As, Sb, and Bi |
| $5X_2 + 2P \rightarrow 2PX_5$ | With excess X$_2$ but not with I$_2$; SbF$_5$, SbCl$_5$, AsF$_5$, AsCl$_5$, and BiF$_5$, may be similarly prepared |
| $X_2 + 2S \rightarrow S_2X_2$ | With Cl$_2$, Br$_2$ |
| $X_2 + H_2O \rightarrow H^+ + X^- + HOX$ | Not with F$_2$ |
| $2X_2 + 2H_2O \rightarrow 4H^+ + 4X^- + O_2$ | F$_2$ rapidly; Cl$_2$, Br$_2$ slowly in sunlight |
| $X_2 + H_2S \rightarrow 2HX + S$ | |
| $X_2 + CO \rightarrow COX_2$ | Cl$_2$, Br$_2$ |
| $X_2 + SO_2 \rightarrow SO_2X_2$ | F$_2$, Cl$_2$ |
| $X_2 + 2X'^- \rightarrow X_2' + 2X^-$ | F$_2$ > Cl$_2$ > Br$_2$ > I$_2$ |
| $X_2 + X_2' \rightarrow 2XX'$ | Formation of the interhalogen compounds (all except iF) |

(d) [11.12]     $Ba^{2+}(aq) + SO_4^{2-}(aq) \rightarrow BaSO_4(s)$

The compounds BaSO$_4$, SrSO$_4$, PbSO$_4$, and Hg$_2$SO$_4$ are fairly insoluble in water; CaSO$_4$ and AgSO$_4$ are slightly soluble. All other known metal sulfates are water solub.e

(e) [11.3]     $3Cl_2(g) + 2Bi(s) \rightarrow 2BiCl_3(s)$   (See Table 11.1.)

(f) [11.3]     $Cl_2(g) + Hg(l) \rightarrow HgCl_2(s)$   (See Table 11.1.)

(g) [11.9]     $Se(s) + Hg(l) \rightarrow HgSe(s)$   (See Table 11.2.)

(h) [11.5]     $CaF_2(s) + H_2SO_4 \xrightarrow{heat} CaSO_4(s) + 2HF(g)$

Hydrogen fluoride is prepared commercially in this way.

(i) [11.5]     $Br_2(g) + H_2(g) \rightarrow 2HBr(l)$

(j) [11.3]     $Br_2(l) + 2Cl^-(aq) \rightarrow NR$   (See Table 11.1.)

TABLE 11.2    Some Reactions of Sulfur, Selenium, and
              Tellurium

| Reaction of Sulfur | Remarks |
|---|---|
| $nS + mM \rightarrow M_mS_n$ | Se, Te react similarly with many metals (not noble metals) |
| $nS + S^{2-} \rightarrow S_{n+1}^{2-}$ | For S and Te, $n = 1$ to 5; for Se, $n = 1$ to 4 |
| $S + H_2 \rightarrow H_2S$ | S > Se > Te; elevated temperatures; compounds are better prepared by actions of dilute HCl on sulfides, selenides, or tellurides |
| $S + O_2 \rightarrow SO_2$ | S > Se > Te; dioxides of Se and Te are easier to prepare with a mixture of $O_2 + NO_2$ |
| $S + 3F_2 \rightarrow SF_6$ | S, Se, Te with excess $F_2$ |
| $S + 2F_2 \rightarrow SF_4$ | S, Se; $TeF_4$ is made indirectly ($TeF_6 + Te$) |
| $2S + X_2 \rightarrow S_2X_2$ | S, Se; $X_2 = Cl_2$ or $Br_2$ |
| $S + 2X_2 \rightarrow SX_4$ | S, Se, Te with excess $Cl_2$; Se, Te with excess $Br_2$; Te with excess $I_2$ |
| $S_2Cl_2 + Cl_2 \rightarrow 2SCl_2$ | $SCl_2$ only, $SBr_2$ unknown; $SeCl_2$, $SeBr_2$ (only in vapor state); $TeCl_2$, $TeBr_2$ are made by thermal decomposition of higher halides |
| $S + 4HNO_3 \rightarrow SO_2 + 4NO_2 + 2H_2O$ | Hot, concentrated nitric acid; S yields mixtures of $SO_2$ and $SO_4^{2-}$; Se yields $H_2SeO_3$ ($SeO_2 \cdot H_2O$); Te yields $2TeO_2 \cdot HNO_3$ |

7.  [4.3, 4.4]    (a)                        (b)

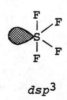

$dsp^3$                    $d^2sp^3$

8. [4.2, 4.3]

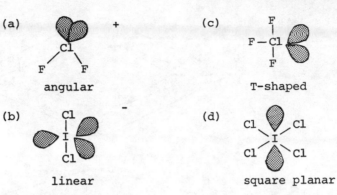

(a) angular    +

(c) T-shaped

(b) linear    −

(d) square planar

9. [11.1]    These gases are obtained by fractional distillation of liquefied air.

10. [11.1]    Helium is obtained from natural gas deposits.

11. [11.1]    The compounds $XeF_3$ and $XeF_5$ would each have an odd number of valence electrons. Most molecules that are easily prepared have an even number of valence electrons.

12. [11.1]    The atoms of the elements of group 0 increase in size from He to Xe. The valence electrons are also less tightly held from He to Xe; i.e., the ionization potentials decrease down the group.

13. [4.5]    See Figure 11.2 of the study guide for the molecular orbital diagram of $F_2$.

14. [8.10, 11.5]    The bond of HF is stronger than that of any other hydrogen halide; therefore, HF is less easily dissociated. When dissociation does occur, $HF_2^-$(aq) species form. Hydrogen bonding stabilizes HF in solution.

15. [11.8]

| Element | Color | Physical State at STP |
|---|---|---|
| Oxygen | Colorless | Gas consisting of $O_2$ molecules |
| Sulfur | Yellow | Solid consisting of $S_8$ rings |
| Selenium | Red to black | Solid consisting of $Se_8$ rings and/or $Se_n$ chains |
| Tellurium | Silver to white | Solid consisting of $Te_n$ chains |

16. [2.15, 2.16]    Te: $1s^2 2s^2 2p^6 3s^2 3p^6 3d^{10} 4s^2 4p^6 4d^{10} 5s^2 5p^4$

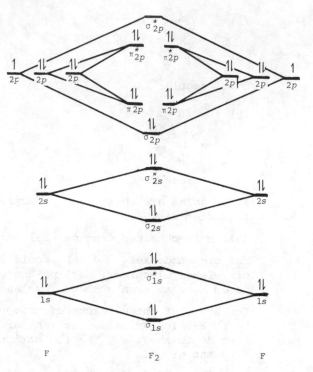

FIGURE 11.2   Molecular orbital diagram of $F_2$

17. [11.9]                    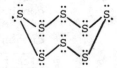         or

18. (a) [11.11]    $H_2SeO_3$, selenous acid
    (b) [11.7]     $HBrO_4$, perbromic acid
    (c) [11.11]    $TeO_2$, tellurium dioxide
    (d) [11.1]     $XeO_4$, xenon(VIII) oxide
    (e) [11.6]     $CaF_2$, calcium fluoride
    (f) [11.12]    $NaHSO_4$, sodium bisulfate
    (g) [11.12]    $H_2S_2O_7$, pyrosulfuric acid
    (h) [11.7]     $HOI$, hypoiodous acid
    (i) [11.6]     $PCl_3$, phosphorus trichloride
    (j) [11.12]    $H_2S_2O_8$, peroxydisulfuric acid
    (k) [11.1]     $Na_4XeO_6$, sodium perxenate
    (l) [11.12]    $Na_2S_2O_3$, sodium thiosulfate

19. [3.10]    Write the formula of the compound in the space provided:

Oxidation

| | Number | Name | Formula |
|---|---|---|---|
| (a) | +1 | hypoiodous acid | _____ |
| (b) | +7 | iodine heptafluoride | _____ |
| (c) | 0 | iodine | _____ |
| (d) | -1 | sodium iodide | _____ |
| (e) | +5 | diiodine pentoxide | _____ |
| (f) | +7 | periodic acid | _____ |
| (g) | -1 | silver(I) iodide or | _____ |
| | | silver iodide | _____ |
| (h) | +3 | iodine trichloride | _____ |
| (i) | +4 | diiodine tetraoxide | _____ |
| (j) | -1 | diarsenic tetraiodide | _____ |
| (k) | +6 | diiodine hexachloride | _____ |

20. [11.12]    The thiosulfate anion has the same structure as the sulfate ion except that one oxygen atom is replaced by a sulfur atom:

21. 3.56 g    Calculate the number of liters of $Cl_2$ at STP:
   [6.9, 11.3]
$$? \text{ liter } Cl_2 = 1.00 \text{ liter}\left(\frac{273 \text{ K}}{298 \text{ K}}\right) = 0.9161 \text{ liter}$$

Calculate the number of grams of $MnO_2$:

$$? \text{ g } MnO_2 = 0.9161 \text{ liter } Cl_2\left(\frac{1 \text{ mol } Cl_2}{22.4 \text{ liter } Cl_2}\right)\left(\frac{1 \text{ mol } MnO_2}{1 \text{ mol } Cl_2}\right)$$

$$\times \frac{86.94 \text{ g } MnO_2}{1 \text{ mol } MnO_2}$$

$$= 3.56 \text{ g } MnO_2$$

22. [3.7, 3.8,    The Lewis structure is
     3.9, 11.7]

From a consideration of electron pair repulsions, we predict the ion to be octahedral:

23. [5.3]    First we determine the molecular weight of DDT:

> 14 carbon atoms:   14(12.011) = 168.15
>  9 hydrogen atoms:  9(1.0079) =   9.0711
>  5 chlorine atoms:  5(35.453) = 177.26
> molecular weight of DDT = 354.48

Then we determine the percentage of chlorine in DDT:

$$\% \ Cl = \left(\frac{177.26 \text{ g/mol DDT}}{354.48 \text{ g/mol DDT}}\right)100$$

$$= 50.006$$

## SELF-TEST

I. We suggest that you go through your lecture notes and
compile a list of chemical reactions, formulas, and names
of compounds. Check your material to make sure that you
did not copy it incorrectly, and then prepare your own
self-test. Write a series of incomplete reactions with
either the reactants or products given and a list of
formulas without names or vice versa. The next day complete the reactions and give either the name or the
formula of the compound without referring to any material.

If your instructor has emphasized the necessity of learning all chemical reactions in Chapter 11 of your text,
complete the following test in 15 minutes.

1. Complete and balance each of the following equations.
   If no reaction occurs, write *NR*.

   (a) $OH^-(aq) + Cl_2(aq) \rightarrow$

   (b) $FeS(s) + H^+(aq) + Cl^-(aq) \rightarrow$

   (c) $Br_2(g) + Ag_2O(s) + H_2O(l) \rightarrow$

   (d) $PCl_3(l) + H_2O(l) \rightarrow$

   (e) $Cl^-(aq) + H_2O(l) \xrightarrow[\text{heat}]{\text{electrolysis}}$

   (f) $Cl^-(aq) + H^+(aq) + H_2O(l) \rightarrow$

   (g) $Cl_2(g) + I^-(aq) \rightarrow$

   (h) $Ba^{2+}(aq) + SO_4^{2-}(aq) \rightarrow$

   (i) $C_{12}H_{22}O_{11}(s) + H_2SO_4(\text{conc.}) \rightarrow$
       sucrose

   (j) $S_2O_3^{2-}(aq) + H^+(aq) \rightarrow$

# The Nonmetals, Part II

OBJECTIVES

I. You should be familiar with the physical properties of the elements of groups III A, IV A, and V A. These elements are shown in the shaded area of the periodic table of Figure 12.1 of the study guide.

II. You should be familiar with the abundance of these elements in nature, the ways in which they can be obtained in pure form, and their uses.

III. You should know the fundamental chemistry of these elements, i.e., the types of compounds they form, how such compounds are prepared, and how such compounds react.

IV. You should be able to apply the basic information of all previous chapters of your text toward an understanding of the chemistry of these elements.

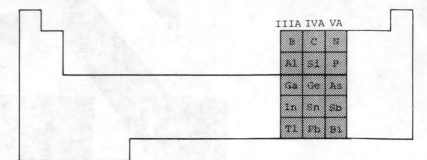

FIGURE 12.1   Periodic table

**EXERCISES**   I. Answer the following:

1. What are the color and physical state of each element of group V A at STP?

2. Write the electronic configuration of each of the following:
   (a) P                    (d) $As^{3+}$
   (b) $P^{3-}$              (e) $As^{5+}$
   (c) As

3. Which element of group V A is most abundant on earth?

4. What is the oxidation number of nitrogen in each of the following compounds? Name each compound.
   (a) $N_2O_4$
   (b) NO
   (c) $N_2O$
   (d) $N_2O_5$
   (e) $N_2O_3$

5. Write the formula of each of the following in the space provided:
   (a) ammonia
   (b) aluminum carbide
   (c) hydroxylamine
   (d) antimony(III) sulfide
   (e) calcium nitride
   (f) phosphorus(V) oxide
   (g) triphosphoric acid
   (h) calcium cyanamide
   (i) phosphorus(III) oxide
   (j) hydrazine
   (k) tetranitrogen tetrasulfide
   (l) phosphine
   (m) nitric oxide

6. Complete and balance the following equations. If no
   reaction occurs, write *NR* in the space provided for
   the products.

   (a) $PCl_3(l) + Cl_2(g) \rightarrow$

   (b) $NaN_3(s) \xrightarrow{\text{heat}}$

   (c) $CaCO_3(s) \xrightarrow{\text{heat}}$

   (d) $NH_3(g) + O_2(g) \xrightarrow[\text{Pt}]{\text{heat}}$

   (e) $P_4O_{10}(s) + H_2O(l) \rightarrow$

   (f) $Hg(CN)_2(s) \xrightarrow{\text{heat}}$

   (g) $CaC_2(s) + H_2O(l) \rightarrow$

   (h) $Sb_2S_2(s) + O_2(g) \rightarrow$

   (i) $NH_4NO_3(s) \xrightarrow{\text{heat}}$

   (j) $AsCl_3(l) + H_2O(l) \rightarrow$

   (k) $Sb(s) + O_2(g) \xrightarrow{\text{heat}}$

   (l) $PCl_5(s) + Cl_2(g) \rightarrow$

   (m) $B(s) + N_2(g) \xrightarrow{\text{heat}}$

7. Predict the structure of $NO_3^-$. Draw the resonance forms
   of the ion and include the formal charges.

8. Given the following electrode potential diagram for
   common nitrogen compounds in acid solution, predict
   which compounds disproportionate:

$$NO_3^- \xrightarrow{+0.94 \text{ V}} HNO_2 \xrightarrow{+1.00 \text{ V}} NO \xrightarrow{+1.59 \text{ V}} N_2O \xrightarrow{+1.77 \text{ V}} N_2 \xrightarrow{+0.27 \text{ V}} NH_4^+$$

(with $+0.96$ V spanning $NO_3^-$ to NO, $+1.12$ V spanning $NO_3^-$ to $HNO_2$ region, and lower span from NO to $N_2$)

9. Draw the Lewis structure of each of the following
   compounds. Predict the shape of each.
   (a) $BF_3$                         (b) $BF_4^-$

10. List two homopolar molecules that are isoelectronic
    with the cyanide ion.

11. Draw the Lewis structure of cyanogen.

12. Which element of group IV A has the highest melting
    point?

13. How many grams of water must be added to 1.00 g $P_4O_{10}$
    to produce tetrametaphosphoric acid?

$$P_4O_{10}(s) + 2H_2O(l) \xrightarrow{0°C} H_4P_4)_{10}(s)$$

14. Which of the following compounds is most basic?
    (a) $PH_3$                    (c) $SbH_3$
    (b) $AsH_3$                   (d) $BiH_3$

15. Which of the following solids is the best electrical conductor?
    (a) $W_2N$                    (c) $BN$
    (b) $S_4N_4$                  (d) $AlN$

16. Which of the following elements forms a three-center bond in some of its compounds?
    (a) Te                        (c) Si
    (b) As                        (d) B

17. Which of the following would not oxidize when heated in air?
    (a) CO                        (c) $N_2$
    (b) $P_4$                     (d) C

18. Which of the following compounds has the lowest boiling point?
    (a) $NH_3$                    (c) $AsH_3$
    (b) $PH_3$                    (d) $SbH_3$

## ANSWERS TO EXERCISES

I. Properties of nonmetals

1. [12.1]

| Element | Color | Physical State at STP |
|---------|-------|----------------------|
| Nitrogen | Colorless | Gas consisting of $N_2$ molecules |
| Phosphorus | White, red, black | Solid consisting of $P_4$ (white) or $P_n$ (black) |
| Arsenic | Gray metallic, yellow | Solid consisting of $As_4$ (yellow) or $As_n$ (gray metallic) |
| Antimony | Gray metallic, yellow | Solid consisting of $Sb_4$ (yellow) or $Sb_n$ (gray metallic |
| Bismuth | Gray metallic | Solid consisting of $Bi_n$ |

2. [12.1]
   (a)   P: $1s^2 2s^2 2p^6 3s^2 3p^3$
   (b)   $P^{3-}$: $1s^2 2s^2 2p^6 3s^2 3p^6$
   (c)   As: $1s^2 2s^2 2p^6 3s^2 3p^6 3d^{10} 4s^2 4p^3$
   (d)   $As^{3+}$: $1s^2 2s^2 2p^6 3s^2 3p^6 3d^{10} 4s^2$
   (e)   $As^{5+}$: $1s^2 2s^2 2p^6 3s^2 3p^6 3d^{10}$

3. nitrogen
    [12.2]

4. [12.2]          After checking the answers, write the formulas of the com-
                   pound in the space provided:

| | Oxidation Number | Name | Formula |
|---|---|---|---|
| (a) | +4 | dinitrogen tetraoxide | |
| (b) | +2 | mononitrogen monoxide, or nitric oxide | |
| (c) | +1 | dinitrogen monoxide, or nitrous oxide | |
| (d) | +5 | dinitrogen pentoxide | |
| (e) | +3 | dinitrogen trioxide | |

5.                After checking the answers, write the name of the com-
                  pound in the space provided:

(a) [12.4]    $NH_3$, _____
(b) [12.11]   $Al_4C_3$, _____
(c) [12.4]    $NH_2OH$, _____
(d) [12.2]    $Sb_2S_3$, _____
(e) [12.3]    $Ca_3N_2$, _____
(f) [12.8]    $P_4O_{10}$, _____
(g) [12.8]    $H_5P_3O_{10}$, _____
(h) [12.4]    $CaNCN$, _____
(i) [12.8]    $P_4O_6$, _____
(j) [12.4]    $N_2H_4$, _____
(k) [12.6]    $H_4S_4$, _____
(l) [12.4]    $PH_3$, _____
(m) [12.4]    $NO$, _____

6. (a) [12.5]    $PCl_3(l) + Cl_2(g) \rightarrow PCl_5(s)$

(b) [12.5]    $2NaN_3(s) \xrightarrow{heat} 2Na(l) + 3N_2(g)$

(c) [12.12]   $CaCO_3(s) \xrightarrow{heat} CaCO(s) + CO_2(g)$

(d) [12.4]    $4NH_3(g) + 5O_2(g) \xrightarrow[Pt]{heat} 4NO(g) + 6H_2O(g)$

(e) [12.8]    $P_4O_{10}(s) + 6H_2O(l) \rightarrow 4H_3PO_4(aq)$

(f) [12.14]   $Hg(CN)_2(s) \xrightarrow{heat} Hg(l) + C_2N_2(g)$

(g) [12.11]   $CaC_2(s) + 2H_2O(l) \rightarrow Ca(OH)_2(aq) + C_2H_2(g)$

(h) [12.2]    $2Sb_2S_3(s) + 9O_2 \xrightarrow{heat} Sb_4O_6(g) + 6SO_2(g)$

(i) [12.7]    $NH_4NO_3 \xrightarrow{heat} N_2O(g) + 2H_2O(g)$

(j) [12.5]    $AsCl_3(l) + 3H_2O(l) \rightarrow H_3AsO_3(aq) + 3H^+(aq) + 3Cl^-(aq)$

(k) [12.8]    $4Sb(s) + 3O_2(g) \xrightarrow{heat} Sb_4O_6(s)$

(l) [12.5]    $PCl_5(s) + Cl_2(g) \rightarrow NR$

(m) [12.17]   $2B(s) + N_2(g) \xrightarrow{heat} 2BN(s)$

7. [12.7]    We predict the molecule to be triangular planar. The resonance structures are

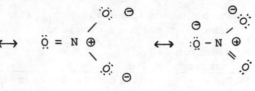

8. [12.7]    $NO_3^-$: stable

$HNO_2$: disproportionates:

$$3HNO_2 \rightarrow 2NO + H_2O + NO_3^- + H^+ \qquad \mathscr{E}° = +0.06 \text{ V}$$

NO: disproportionates

$$4NO + H_2O \rightarrow N_2O + 2HNO_2 \qquad \mathscr{E}° = +0.59 \text{ V}$$

Write another disproportionation reaction giving stable products.

$N_2O$: disproportionates

$$5N_2O + H_2O \rightarrow 4N_2 + 2NO_3^- + 2H^+ \qquad \mathscr{E}° = +0.65 \text{ V}$$

$N_2$: stable

$NH_4^+$: stable

9. [12.17]

(a)     :F:
        |
        B
       / \
    F      F

triangular
planar

        :F:        ⁻
        |
    :F—B—F:
        |
        :F:

tetrahedral

10. $N_2$, CO
    [12.14]

11. :N≡C—C≡N:
    [12.14]

12. carbon
    [12.9]

13. 0.127 g

We calculate the number of grams of water:

$$? \text{ g } H_2O = 1.00 \text{ g } P_4O_{10}\left(\frac{1 \text{ mol } P_4O_{10}}{283.9 \text{ g } P_4O_{10}}\right)\left(\frac{2 \text{ mol } H_2O}{1 \text{ mol } P_4O_{10}}\right)$$

$$\times \left(\frac{18.02 \text{ g } H_2O}{1 \text{ mol } H_2O}\right)$$

$$= 0.127 \text{ g } H_2O$$

14. $PH_3$
    [12.4]

15. $W_2N$
    [12.3]

Ionic nitrides are better conductors than covalent nitrides.

16. B
    [12.16]

17. $N_2$
    [8.1]

Air is composed of 21 percent and 78 percent nitrogen by volume. Oxygen and nitrogen do not react with each other.

18. $PH_3$
    [8.9]

## SELF-TEST

I. We suggest that you go through your lecture notes and compile a list of chemical reactions, formulas, and names of compounds. Check your material to make sure that you did not copy any material incorrectly and then prepare your own self-test. Write a series of incomplete reactions with either the reactants or products given and a list of formulas without names or vice versa. The next day complete the reactions and give either the name or the formula of the compound without referring to any material. You may wish to exchange tests with your friends.

If your instructor has emphasized the necessity of learning all chemical reactions in Chapter 12 of your text, complete the following test in 10 minutes:

1. Draw the resonance structures of nitric acid.

2. Complete the following reactions:
   (a) $Sb_2S_3(s) + O_2(g) \rightarrow$
   (b) $CaO(s) + C(s) \rightarrow$
   (c) $Ca_3N_2(s) + H_2O(l) \rightarrow$
   (d) $As_4O_6(s) + C(s) \rightarrow$
   (e) $PBr_3(l) + H_2O(l) \rightarrow$
   (f) $PBr_3(l) + O_2(g) \rightarrow$
   (g) $PI_3(s) + I_2(s) \rightarrow$

3. Hypophosphorous acid is a
   (a) monoprotic acid          (c) triprotic acid
   (b) diprotic acid            (d) base

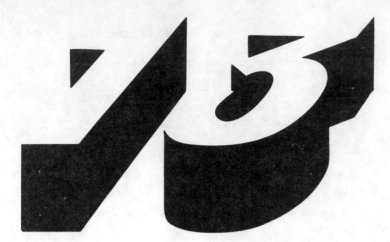

# Chemical Kinetics and Chemical Equilibrium

OBJECTIVES

I. You should be able to demonstrate your knowledge of the following terms by defining them, describing them, or giving specific examples of them:

activated complex [13.3]
adsorption [13.7]
Arrhenius equation [13.6]
catalyst [13.7]
chain mechanism [13.5]
chemisorption [13.7]
collision theory [13.3]
energy of activation [13.3]
equilibrium constant [13.8, 13.9, 13.10]
frequency factor [13.6]
initial rate [13.1]
Le Chatelier's principle [13.11]
molecularity [13.4]
order [13.2]

rate constant [13.2]
rate-determining step [13.5]
reaction intermediates [13.3, 13.7]
reaction mechanism [13.5]
reaction rate [13.1]
single step [13.3]
steady-state [13.5]
transition state [13.3]

II. Given experimental data, you should be able to formulate the rate equation for a reaction.

III. You should be able to propose a reaction mechanism based on experimental data.

IV. You should be able to use the Arrhenius equation to calculate $E_a$ or $k$ from appropriate experimental data.

V. You should understand the effect that a change in conditions such as temperature and concentration will have on the reaction rate, as well as the effect that the presence of a catalyst will have.

VI. Given a chemical equation, you should be able to write the expression of the equilibrium constant of the reaction.

VII. Given the concentrations of species at equilibrium, you should be able to evaluate the equilibrium constant, $K$, for a reaction.

VIII. You should be able to use an equilibrium constant to determine the concentration of a species at equilibrium.

IX. You should be able to use Le Chatelier's principle to determine the effect that a change in conditions will have on an equilibrium system.

EXERCISES

I. Answer each of the following with *true* or *false*. If a statement is false, correct it.

_____ 1. The reaction represented in the energy diagram of Figure 13.1 of the study guide is exothermic.

_____ 2. In the energy diagram of Figure 13.1 of the study guide, $E_1$ is the activation energy for the forward reaction.

_____ 3. In the energy diagram of Figure 13.1 of the study guide, $E_2$ is $\Delta E$ for the reaction.

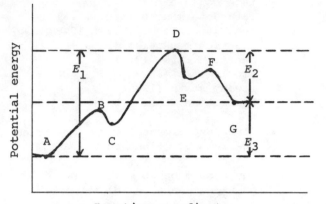

FIGURE 13.1  Energy diagram

_____ 4. Point A in Figure 13.1 represents the potential energy of all reactants.

_____ 5. Point B in Figure 13.1 represents the potential energy of a reaction intermediate.

_____ 6. The energy diagram, Figure 13.1, could be only for a catalyzed reaction.

_____ 7. If the specific rate constant for a reaction is large, the reaction will proceed rapidly.

_____ 8. Rate constants are independent of temperature.

_____ 9. Reaction intermediates appear in the net equation of a reaction.

_____ 10. A catalyst changes the path of a reaction.

_____ 11. A catalyst changes the position of equilibrium.

_____ 12. If the equilibrium constant for a reaction is large, the reaction will proceed rapidly.

_____ 13. For the reaction

$$3Fe(s) + 4H_2O(g) \rightarrow Fe_3O_4(s) + 4H_2(g)$$

the expression for the equilibrium constant is

$$K = \frac{[H_2]^4}{[H_2O]^4}$$

_____ 14. Equilibrium constants are not affected by pressure.

_____ 15. When a system reaches a state of dynamic equilibrium, the reaction stops.

II. Do the following:

1. For the reaction

   A(g) + B(g) → C(g)

   the following data were obtained from three experiments:

   | Experiment | [A] | [B] | Rate (Formation of C) |
   |---|---|---|---|
   | 1 | 0.20M | 0.20M | $3.0 \times 10^{-4} M$/min |
   | 2 | 0.60M | 0.60M | $81.0 \times 10^{-4} M$/min |
   | 3 | 0.60M | 0.20M | $9.0 \times 10^{-4} M$/min |

   Use the data to answer the following:

   (a) What is the rate equation for the reaction?
   (b) What is the order of the reaction?
   (c) What is the numerical value of the specific rate constant?
   (d) If the volume of the reaction vessel were halved, what would be the effect on the reaction rate?
   (e) Is the reaction as written a one-step process?

2. Assume that the rate-determining step of a reaction is

   2A(g) + B(g) → C(g)

   and that 2 mol of A and 1 mol of B are mixed in a liter container. Compare the following with the initial reaction rate of the mixture:
   (a) rate when half of A has been consumed
   (b) rate when two-thirds of A has been consumed
   (c) initial reaction rate of a mixture of 2 mol of A and 2 mol of B in a liter container
   (d) initial reaction rate of a mixture of 4 mol of A and 2 mol of B in a liter container

3. For the reaction

   $H_2$(g) + $I_2$(g) → 2HI(g)

   the slope of the plot of log $k$ vs. $1/T$ is -8880 K. What is the energy of activation for the reaction? (Figure 13.10 of your text gives similar data for another reaction.)

4. For the reaction

   A(g) + B(g) → C(g)

   the specific rate constants of the forward and backward reactions are $0.16 M^{-1}$ sec$^{-1}$ and $4.0 \times 10^4$ sec$^{-1}$, respectively. If each reaction occurs in one step, what is the value of the equilibrium constant?

5. Write the expression for the equilibrium constant, $K$, for each of the following reactions:

    (a) $H_2(g) + I_2(g) \rightleftharpoons 2HI(g)$

    (b) $PCl_5(g) \rightleftharpoons PCl_3(g) + Cl_2(g)$

    (c) $N_2(g) + 3H_2(g) \rightleftharpoons 2NH_3(g)$

    (d) $C(s) + CO_2(g) \rightleftharpoons 2CO(g)$

    (e) $2NOBr(g) \rightleftharpoons 2NO(g) + Br_2(g)$

    (f) $2Cl_2(g) + 2H_2O(g) \rightleftharpoons 4HCl(g) + O_2(g)$

    (g) $2Hg(g) + O_2(g) \rightleftharpoons 2HgO(s)$

6. A 1.00-liter vessel initially contains 2.01 mol of $N_2$ and 2.08 mol of $H_2$ at 500°C. After the system reaches equilibrium, 0.50 mol of $NH_3$ has formed. Answer the following:

    (a) What are the equilibrium concentrations of $N_2$ and $H_2$?

    (b) What is the equilibrium constant, $K$, at 500°C?

    (c) What is $K_p$ at 500°C?

7. For the equilibrium

    $A(g) + B(g) \rightleftharpoons C(g)$

    $K$ is $1.00 \times 10^{-3}$ liter/mol at 100°C and $1.00 \times 10^{-7}$ liter/mol at 200°C. Is the reaction endothermic or exothermic?

8. At temperature $T_1$ and a total pressure of 1.00 atm, $N_2O_4$ is 20.0 percent dissociated:

    $N_2O_4(g) \rightleftharpoons 2NO_2(g)$

    Assume that 1.00 of $N_2O_4$ is present initially and answer the following:

    (a) How many moles of $N_2O_4(g)$ and $NO_2(g)$ are present at equilibrium?

    (b) What is the total number of moles of gas present at equilibrium?

    (c) What are the equilibrium pressures of $N_2O_4(g)$ and $NO_2(g)$?

    (d) What is the value of $K_p$ at temperature $T_1$?

9. At 100°C the equilibrium constant, $K$, for the reaction

    $CO(g) + Cl_2(g) \rightleftharpoons COCl_2(g)$

    is $4.57 \times 10^9$ liter/mol. If 1.00 mol of $COCl_2$ is placed in a 1.00-liter container, what is the concentration of CO after equilibrium has been established? (Note that the subtraction of a very small number from a large number may be neglected.)

10. The equilibrium constant, $K_p$, for the reaction

$$C_4H_{10}(g) \rightleftarrows 2H_2(g) + C_4H_6(g)$$

is $1.0 \times 10^{-6}$ atm at 600°C. If 2.0 mol of $H_2$ and 1.0 mol of $C_4H_{10}$ are put into a 1.00-liter flask at 600°C, how many moles of $C_4H_6$ will be formed?

11. Consider the equilibrium

$$2SO_2(g) + O_2(g) \rightleftarrows 2SO_3(g) + \text{heat}$$

What effect would each of the following stresses have on the reaction?
(a) addition of $SO_2$ at constant $V$ and $T$
(b) reduction of volume at constant temperature
(c) increase in temperature

12. At 400°C equilibrium for the reaction

$$27 \text{ kcal} + 2Cl_2(g) + 2H_2O(g) \rightleftarrows 4HCl(g) + O_2(g)$$

is established. What would be the effect on the number of moles of $Cl_2(g)$ present if each of the following changes were made?
(a) increase of temperature to 600°C
(b) addition of $O_2(g)$
(c) removal of $H_2O(g)$
(d) reduction of the volume of the container
(e) introduction of a catalyst

13. What is the concentration of HI at equilibrium at 357°C if $6.22M$ $H_2$ and $5.71M$ $I_2$ are in a container and $K$ for the reaction

$$H_2(g) + I_2(g) \rightleftarrows 2HI(g)$$

is 71.3?

ANSWERS TO EXERCISES

I. Principles of chemical kinetics and equilibrium

1. False
   [13.5]

The reaction is endothermic. The energy of the products is higher than that of the reactants; thus, the addition of energy is necessary for the reaction to occur.

2. True
   [13.5]

3. False
   [13.5]

Since $E_2$ is the difference between the energy of the products and that of the activated complex, it is the activation energy for the backward reaction. The $\Delta E$ for the forward reaction is $E_3$, the difference between the energy of the products and that of the reactants.

4. True
   [13.5]

5. False      This is the potential energy of an activated complex.
   [13.5]     Intermediates are at points C and E in Figure 13.1 of
              the study guide.

6. False      You cannot tell from a single diagram whether a catalyst
   [13.5, 13.7]  is present or not.

7. True
   [13.2]

8. False      The rate constant, $k$, varies with temperature, $T$, in a
   [13.6]     manner described by the Arrhenius equation

              $$k = Ae^{-E_a/RT}$$

9. False      Reaction intermediates are not final products; thus,
   [13.7]     they would not appear in the net equation.

10. True
    [13.7]

11. False      A catalyst has no effect on the position of equilibrium
    [13.7, 13.10]  since it affects the rate of the forward reaction and
               that of the backward reaction to an equal extent.

12. False      The equilibrium constant indicates nothing about how a
    [13.8]     reaction will occur. It only gives a ratio of products
               to reactants.

13. True
    [13.9]

14. True
    [13.10]

15. False      In a state of dynamic equilibrium the reaction continues,
    [13.8]     but the forward and reverse rates are equal.

## II. Chemical kinetics and equilibrium calculations

1. (a) rate      Comparing experiments 1 and 3, we find
   $= k[A][B]^2$
   [13.1,            [B] = constant
   13.2]
                     [A] changes

                 rate $= k[A]^x[B]^y$

   Since $k$ and, in this case, $[B]^y$ are constants, the rate
   is proportion to $[A]^x$. In experiment 3, [A] is 3 times
   greater than than in experiment 1. The rate in experiment 3 is also 3 times greater than that in experiment 1.
   The rate is directly proportional to the change in [A];
   thus,

                 $x = 1$

and

$$rate = k[A][B]^y$$

Comparing experiments 2 and 3 we find

[A] = constant

[B] changes

$$rate = k[A]^x[B]^y$$

Since $k$ and, in this case, $[A]^x$ are constants, the
rate is proportional to $[B]^y$. In experiment 2, [B] is
3 times greater than that in experiment. 3. The rate
in experiment 2 is 9 times greater than that in experiment 3. The rate is directly proportional to the square
of the change in [B]; thus,

$$y = 2$$

and

$$rate = k[A][B]^2$$

(b) 3
[13.2]

The reaction order is the sum of the exponents of the
concentrations appearing in the equation. Thus,

$$\text{reaction order} = \text{sum of exponents}$$
$$= 1 + 2$$
$$= 3$$

(c) $0.038M^{-2}$ min$^{-1}$

We can use any set of data in the equation

$$rate = k[A][B]^2$$

Using the data from experiment 1, we find

$$= \frac{rate}{[A][B]^2}$$

$$= \frac{3.0 \times 10^{-4} /min}{(0.20M)(0.20M)^2}$$

$$= 3.8 \times 10^{-2} \text{ liter}^2 \text{ mol}^{-2} \text{ min}^{-1}$$

$$3.8 \times 10^{-2} M^{-2} \text{ min}^{-1}$$

(d) The rate
would be 8
times
greater
than that
of the
original.

The rate equation is

$$rate = k[A][B]^2$$

If the volume of the reaction vessel were halved, all
concentrations would be doubled. For example, if the
concentrations were originally 1 mol/1 liter, they
would be 1 mol/0.5 liter, or 2M, when the volume
were halved. Thus, the rate of the reaction when the
volume were halved would be 8 times greater than that
of the original:

$$\text{rate} = [A][B]^2$$

$$= (2M)(2M)^2$$

$$= (8M^3)$$

(e) no
[13.3]

If this were a one-step process, the reaction as written would be the rate-determining step and the exponents of the concentrations of the reactants in the rate equation would be the same as the coefficients in the chemical reaction:

$$\text{rate} = [A]^1[B]^1$$

2. (a) one-eighth of the original rate
[13.2]

The rate-determining step is

$$2A(g) + B(g) \rightarrow C(g)$$

Thus,

$$\text{rate} = k[A]^2[B]$$

and initially

$$\text{rate} = k[A]^2[B]$$

$$= k(2M)^2(1M)$$

$$= k(4M^3)$$

When half of each reactant has been consumed,

$$\text{rate} = k[A]^2[B]$$

$$= k(1M)^2(0.5M)$$

$$= k(0.5M^3)$$

Thus, the rate when half of each reactant has been consumed is one-eighth of the initial rate.

(b) one-twenty-seventh of the initial rate
[13.2]

When two-thirds of A has been consumed, one-third of A remains $(1/3) \times 2 = (2/3)$. Likewise, one-third of B remains:

$$\text{rate} = [A]^2[B]$$

$$= k(2/3 \ M)^2(1/3 \ M)$$

$$= k(4/9 \ M^2)(1/3 \ M)$$

$$= k(4/27 \ M^3)$$

$$= k(4M^3)(1/27)$$

Thus, the rate when two-thirds of A has been consumed is one-twenty-seventh of the initial rate.

(c) The initial rate of the mixture of 2 mol of A and 2 mol of B is twice as great as that of the mixture of 2 mol of A and 1 mol of B.
[13.2]

The initial rate of a mixture of 2 mol of A and 2 mol of B in a liter container is

$$rate = k[A]^2[B]$$
$$= k(2M)^2(2M)$$
$$= k(8M^3)$$
$$= k(4M^3)(2)$$

Thus, the initial rate of a mixture of 2 mol of A and 2 mol of B in a liter container is 2 times as great as that of a mixture of 2 mol of A and 1 mol of B in a liter container.

(d) The initial rate of the mixture of 4 mol of A and 2 mol of B is 8 times as great as that of the mixture of 2 mol of A and 1 mol of B.
[13.2]

The initial rate of a mixture of 4 mol of A and 2 mol of B in a liter container is

$$rate = k[A]^2[B]$$
$$= k(4M)^2(2M)$$
$$= k(32M^3)$$
$$= k(4M^3)(8)$$

Thus, the initial rate of a mixture of 4 mol of A and 2 mol of B in a liter container is 8 times as great as the initial rate of a mixture of 2 mol of A and 1 mol of B in a liter container.

3. 169.9 kJ/mol
[13.6]

The Arrhenius equation is

$$k = Ae^{-E_a/RT}$$

If we take the natural log of the Arrhenius equation, we find

$$\ln k = \ln A - \left(\frac{E_a}{RT}\right)$$

If we change the preceding equation to common logs, we find

$$2.303 \log k = 2.303 \log A - \left(\frac{E_a}{RT}\right)$$

$$\log k = \log A - \frac{E_a}{2.303RT}$$

$$\log = \frac{-E_a}{2.303RT} + \log A$$

The preceding equation is the equation of a straight line, $y = mx + b$. The slope of a plot of $\log k$ vs. $1/T$ is $-E_a/2.303R$. Thus,

$$m = -\frac{E_a}{2.303R}$$

$$E_a = -2.303Rm$$

$$= (-2.303)(8.314 \text{ J/K mol})(-8880 \text{ K})$$

$$= 169.9 \text{ kJ}$$

4. $4.0 \times 10^{-6} M^{-1}$
   [13.8]

The relationship between the equilibrium constant and the rates of the forward and backward reactions is

$$K = \frac{k_f}{k_b}$$

$$= \frac{0.16 M^{-1} \text{ sec}^{-1}}{4.0 \times 10^4 \text{ sec}^{-1}}$$

$$= 4.0 \times 10^{-6} M^{-1}$$

5. [13.9]

The exponent of each species is equal to the coefficient of that species in the equation of the chemical reaction.

(a) $K = \dfrac{[HI]^2}{[H_2][I_2]}$

(b) $K = \dfrac{[PCl_3][Cl_2]}{[PCl_5]}$

(c) $K = \dfrac{[NH_3]^2}{[N_2][H_2]^3}$

(d) $K = \dfrac{[CO]^2}{[CO_2]}$    The concentrations of solids at constant temperature are constant and are included in the value of $K$; therefore, they are not written in the expression for the equilibrium constant.

(e) $K = \dfrac{[NO]^2[Br_2]}{[NOBr]^2}$

(f) $K = \dfrac{[HCl]^4[O_2]}{[Cl_2]^2[H_2O]^2}$

$$= \frac{1}{[Hg]^2[O_2]}$$

6. [13.9]

At equilibrium

(a) $[N_2] = 1.76 M$     $[NH_3] = 0.50 \text{ mol}/1.00 \text{ liter} = 0.50 M$

$[H_2] = 1.33 M$    The balanced equation

$$N_2(g) + 3H_2(g) \rightleftharpoons 2NH_3(g)$$

indicates that for every 2 mol of $NH_3$ formed, 1 mol of $N_2$ is lost. Thus, at equilibrium

$$[N_2] = \frac{2.01 \text{ mol } N_2 - (0.50 \text{ mol } NH_3 \text{ formed})\left(\dfrac{1 \text{ mol } N_2 \text{ lost}}{2 \text{ mol } NH_3 \text{ formed}}\right)}{1.00 \text{ liter soln.}}$$

$$= 1.76M$$

Similarly, for every 2 mol of $NH_3$ formed, 3 mol of $H_2$ is lost:

$$[H_2] = \frac{2.08 \text{ mol } H_2 - (0.50 \text{ mol } NH_3 \text{ formed})\left(\dfrac{3 \text{ mol } H_2 \text{ lost}}{2 \text{ mol } NH_3 \text{ formed}}\right)}{1.00 \text{ liter soln.}}$$

$$= 1.33M$$

(b) $K = 6.0 \times 10^{-2}$ $M^{-2}$ [13.9]

We substitute concentrations into the expression for the equilibrium constant:

$$K = \frac{[NH_3]^2}{[N_2][H_2]^3}$$

$$= \frac{(0.50)^2}{(1.76M)(1.33M)^3}$$

$$= 6.0 \times 10^{-2} M^{-2}, \text{ or } 6.0 \times 10^{-2} \text{ liter}^2 \text{ mol}^{-2}$$

(c) $K_p = 1.5 \times 10^{-5}$ $atm^{-2}$ The relationship between $K_p$ and $K$ is

$$K_p = K(RT)^{\Delta n}$$

For the reaction,

$$\Delta n = -2$$

and

$$K_p = K(RT)^{\Delta n}$$

$$= (6.0 \times 10^{-2} \text{ liter}^2 \text{ mol}^{-2})$$
$$\times [(0.0821 \text{ liter atm } °K^{-1} \text{ mol}^{-1})(773 \text{ K})]^{-2}$$

$$= 1.5 \times 10^{-5} \text{ atm}^{-2}$$

7. exothermic [13.10]

The expression for the equilibrium constant is

$$K = \frac{[C]}{[A][B]}$$

The equilibrium constant, $K$, decreases with increasing temperature, a fact that implies that the numerator of the equilibrium expression is becoming smaller. Thus, the equilibrium is shifting to the left. Since an increase in temperature shifts the equilibrium to the left, heat must appear on the right side; therefore, heat is released, and the reaction is exothermic.

8. (a) 0.80 mol $N_2O_4$

   0.40 mol $NO_2$

   [13.10]

The $N_2O_4$ is 20.0 percent dissociated; therefore, 0.80 mol of $N_2O_4$ remains:

? mol $N_2O_4$ remaining = 1.00 mol $N_2O_4$

$$- 0.200(1.00) \text{ mol } N_2O_4$$

$$= 0.80 \text{ mol } N_2O_4$$

Since 0.20 mol of $N_2O_4$ is dissociated, we know from the balanced equation that 0.40 mol of $NO_2$ is formed:

? mol $NO_2$ formed = 0.20 mol $N_2O_4$ lost $\dfrac{2 \text{ mol } NO_2 \text{ formed}}{1 \text{ mol } N_2O_4 \text{ lost}}$

$$= 0.40 \text{ mol } NO_2$$

(b) 1.20 mol gas

We calculate the total number of moles of gas present at equilibrium:

moles of gas = 0.80 mol gas + 0.40 mol gas

$$= 1.20 \text{ mol gas}$$

(c) $p_{N_2O_4} = 0.667$ atm

We determine the partial pressure of each gas:

$$p_{N_2O_4} = \left( \frac{n_{N_2O_4}}{n_{N_2O_4} + n_{NO_2}} \right) p$$

$$= \left( \frac{0.80 \text{ mol}}{0.80 \text{ mol} + 0.40 \text{ mol}} \right) 1.00 \text{ atm}$$

$$= 0.667 \text{ atm}$$

$$p_{NO_2} = \left( \frac{n_{NO_2}}{n_{N_2O_4} + n_{NO_2}} \right) p$$

$$= \left( \frac{0.40 \text{ atm}}{0.80 \text{ atm} + 0.40 \text{ atm}} \right) 1.00 \text{ atm}$$

$$= 0.333 \text{ atm}$$

(d) $1.66 \times 10^{-1}$ atm

We substitute partial pressures into the expression for $K_P$:

$$K_P = \frac{(p_{NO_2})^2}{p_{N_2O_4}}$$

$$= \frac{(0.333 \text{ atm})^2}{(0.667 \text{ atm})}$$

$$= 0.166 \text{ atm}$$

9. $[CO] = 1.48 \times 10^{-5}M$  At 100°C the equilibrium constant is
   [13.10]

$$K = 4.57 \times 10^9 \text{ liter/mol} = \frac{[COCl_2]}{[CO][Cl_2]}$$

We make the following assumptions:

| Species | Original Concentration | Equilibrium Concentration |
|---------|------------------------|---------------------------|
| CO | 0 | $(x)M$ |
| $Cl_2$ | 0 | $(x)M$ |
| $COCl_2$ | 1.00 | $(1.00 - x = 1.00)M$ |

We substitute concentrations into the expression for the equilibrium constant:

$$= 4.57 \times 10^9 \text{ liter/mol} = \frac{[COCl_2]}{[CO][Cl_2]}$$

$$4.57 \times 10^9 \text{ liter/mol} = \frac{1.00}{x^2}$$

$$x^2 = 2.19 \times 10^{-10}M^2$$

$$x = 1.48 \times 10^{-5}M$$

Note that $1.48 \times 10^{-5}M$ is negligible compared with $1.00M$.

10. $4.8 \times 10^{-11}$   The change in the number of moles, $\Delta n$, is +2. Therefore,
    mol $C_4H_6$
    [13.10]

$$K = K_P(RT)^{-\Delta n}$$

$$= 1.0 \times 10^{-6} \text{ atm}^2 [(0.0821 \text{ liter atm K}^{-1} \text{ mol}^{-1})(873 \text{ K})]^{-2}$$

$$= 1.9 \times 10^{-10}M^2$$

The equilibrium expression is

$$K = \frac{[H_2]^2[C_4H_6]}{[C_4H_{10}]}$$

From the reaction equation we know that for every 1 mol of $C_4H_6$ formed, 2 mol of $H_2$ are formed. Also, for every 1 mol of $C_4H_6$ formed, 1 mol of $C_4H_{10}$ is lost. Therefore, we make the following assumptions:

| Species | Original Concentration | Equilibrium Concentration |
|---------|------------------------|---------------------------|
| $C_4H_6$ | 0 | $(x)M$ |
| $H_2$ | 2.0M | $(2.0 + 2x - 2.0)M$ |
| $C_4H_{10}$ | 1.0M | $(1.0 - x = 1.0)M$ |

We substitute concentrations into the expression for the equilibrium constant:

$$K = 1.9 \times 10^{-10}M^2 = \frac{(2.0M)^2(x)}{(1.0M)}$$

$$x = \frac{1.9 \times 10^{-10}\ 2}{4.0M}$$

$$x = 4.8 \times 10^{-11}M$$

Note that $4.8 \times 10^{-11}M$ is negligible compared with $1.0M$ and $9.6 \times 10^{-11}M$ is negligible compared with $2.0M$.

11. [13.11]

   (a) shift to right

   A shift to the right occurs since the stress is on the left.

   (b) shift to right

   A shift to the right occurs since the stress is on the left. The 2 moles of gas on the right would occupy less volume than the 3 moles of gas on the left.

   (c) shift to left

   A shift to the left occurs since the stress is on the right. The reaction is exothermic; thus, the shift is toward the left when the temperature is increased.

12. [13.11]

   (a) decrease

   The number of moles of $Cl_2$ would decrease. The reaction is endothermic, and the increase in temperature would shift the reaction to the side without heat.

   (b) increase

   The number of moles of $Cl_2$ would increase. An increase in the number of moles of $O_2$ would shift the reaction away from the side with $O_2$.

   (c) increase

   The number of moles of $Cl_2$ would increase. The removal of $H_2O$ would cause a shift to the left.

   (d) increase

   The number of moles of $Cl_2$ would increase. There are fewer moles of gas on the left side than on the right. Thus, to alleviate the stress of volume reduction, the equilibrium would shift to the side that contains fewer molecules since such molecules collectively would occupy less volume.

   (e) no change

   There would be no change in the number of moles of $Cl_2$ since a catalyst would have no effect on the position of equilibrium.

13. [HI] = 9.55$M$ [13.9, 13.10]

   The expression for the equilibrium constant is

   $$K = \frac{[HI]^2}{[H_2][I_2]}$$

   From the balanced equation we know that for every 2 mol of HI formed, 1 mol of $H_2$ and 1 mol of $I_2$ are lost, or for every 1 mol of HI formed, $\frac{1}{2}$ mol of $H_2$ and $\frac{1}{2}$ mol of $I_2$ are lost. Therefore, we make the following assumptions:

| Species | Original Concentration | Equilibrium Concentration |
|---------|------------------------|---------------------------|
| HI | 0 | $(x)M$ |
| $H_2$ | 6.22 | $(6.22 - x/2)M$ |
| $I_2$ | 5.71 | $(5.71 - x/2)M$ |

We substitute concentrations into the expression for the equilibrium constant and use the quadratic equation

$$x = \frac{-b \pm \sqrt{b^2 - 4ac}}{2a}$$

to solve for $x$:

$$K = 71.3 = \frac{[HI]^2}{[H_2][I_2]}$$

$$= \frac{x^2}{(6.22 - x/2)(5.71 - x/2)}$$

$$= \frac{x^2}{35.5 - 2.86x - 3.11x + x^2/4}$$

$$x^2 = 2531 - 426x + 17.8x^2$$

$$0 = 16.8x^2 - 426x + 2531$$

$$x = \frac{-(-426) \pm \sqrt{(-426)^2 - 4(16.8)(2531)}}{2(16.8)}$$

$$= 9.55 \text{ or } 15.8$$

Since the concentration of $H_2$ and that of $I_2$ originally present are smaller than $x/2$ when $x = 15.8$, the solution is $x = 9.55$. Therefore,

$$HI = 9.55M$$

SELF-TEST

I. Complete the test in 45 minutes. Answer each of the following:

1. The chemical equation for the reaction between oxalate ion and mercuric chloride in aqueous solution is

$$C_2O_4^{2-}(aq) + 2HgCl_2(aq) \rightarrow 2CO_2(g) + Hg_2Cl_2(s) + 2Cl^-(aq)$$

At temperature    the following kinetic data were obtained:

| Experiment | $[HgCl_2]$ | $[C_2O_4^{2-}]$ | Rate (Formation of $CO_2$) |
|------------|-----------|-----------------|-----------------------------|
| 1 | 0.080M | 0.200M | $1.5 \times 10^{-5}M$ min$^{-1}$ |
| 2 | 0.080M | 0.400M | $6.0 \times 10^{-5}M$ min$^{-1}$ |
| 3 | 0.040M | 0.400M | $3.0 \times 10^{-5}M$ min$^{-1}$ |

Calculate the kinetic order for $HgCl_2$, the kinetic order for $C_2O_4^{2-}$, and the experimental rate constant with the appropriate units.

2. The bromination of acetone in aqueous solution is given by the equation

$$\underset{\text{acetone}}{H_3C-\overset{\overset{\displaystyle O}{\|}}{C}-CH_3}(aq) + Br_2(aq) \rightarrow H_3C-\overset{\overset{\displaystyle O}{\|}}{C}-CH_2Br(aq) + HBr(aq)$$

and has an experimental rate law

rate $= k[\text{acetone}]^1$

Answer each of the following with *increases*, *decreases*, or *remains the same*:

(a) As the concentration of acetone is increased, the experimental rate constant _____.

(b) As the concentration of acetone is increased, the rate _____.

(c) As the concentration of $Br_2$ is increased, the rate _____.

(d) If a suitable positive catalyst is introduced, the overall activation energy for the reaction _____.

(e) If the temperature is increased, the rate constant for the rate-determining step _____.

3. The reaction of gaseous chlorine and water

$$2Cl_2(g) + 2H_2O(g) \rightleftharpoons 4HCl(g) + O_2(g)$$

is endothermic. The four reactant gases are mixed in a reaction vessel and allowed to attain an equilibrium state. Indicate the effect (i.e., *increase*, *decrease*, or *no change*) of the following operations described in the left column on the equilibrium value of the quantity in the right column. Each operation is to be considered separately. Temperature and volume are constant except when the contrary is indicated.

|  | Operation | Quantity |
|---|---|---|
| _____ | (a) increase in volume of container | number of moles of HCl |
| _____ | (b) increase in volume of container | $K_P$ |
| _____ | (c) addition of $O_2(g)$ | $K_P$ |
| _____ | (d) addition of $O_2(g)$ | number of moles of $O_2$ |
| _____ | (e) increase in temperature | $K_P$ |

4. The decomposition of $CaSO_3(s)$

$$CaSO_3(s) \rightleftharpoons CaO(s) + SO_2(g)$$

is endothermic. At equilibrium there are significant quantities of all three compounds in the reaction vessel. Suggest two ways in which the yield of $SO_2(g)$ might be appreciably increased.

5. At 269.9 K nitric oxide and bromine are mixed in a vessel. Initially the partial pressures are

$$p_{NO} = 0.129 \text{ atm}$$

$$p_{Br_2} = 0.0543 \text{ atm}$$

The following equilibrium is established:

$$2NO(g) + Br_2(g) \rightleftharpoons 2NOBr(g)$$

The total pressure at equilibrium is 0.145 atm. Calculate $K_p$.

6. At temperature $T$ 1.00 mol of NOCl is sealed in a 1.00-liter container. At equilibrium 20.0 percent of the NOCl is dissociated:

$$2NOCl(g) \rightleftharpoons 2NO(g) + Cl_2(g)$$

What is the value of the equilibrium constant? Use molar concentrations.

# Elements of Chemical Thermodynamics

**OBJECTIVES**

I. You should be able to demonstrate your knowledge of the following terms by defining them, describing them, or giving specific examples of them:

absolute entropy [14.8]
activity [14.9]
bond energy [14.3]
enthalpy [14.2]
entropy [14.4]
equilibrium [14.5, 14.9]
first law of thermodynamics [14.1]
Gibbs free energy [14.5]
internal energy [14.1]
second law of thermodynamics [14.4]
spontaneous [14.5]
standard enthalpy of formation [14.2]
standard state [14.9]

state function [14.1]
surroundings [14.1]
system [14.1]
third law of thermodynamics [14.8]
work [14.1]

II. You should be able to calculate $\Delta H$ for a reaction from enthalpies of combustion, enthalpies of formation, bomb calorimeter data, and bond dissociation energies.

III. You should be able to calculate $K$, $\Delta G$, and $\Delta S$ of an electrochemical cell reaction from the values of $\Delta H$, $\mathscr{E}$, and $T$.

IV. Using tabulated values of the Gibbs free energy of formation, $\Delta G_f^\circ$, you should be able to calculate $\Delta G^\circ$ of a reaction.

V. You should be able to determine equilibrium constants from standard electrode potential data.

## EXERCISES

I. Answer each of the following with *true* or *false*. If the statement is false, correct it.

_____

1. Consider a 1-ml sample of pure water at 25°C and 1 atm (state A). The sample is cooled to 1°C and then the pressure is reduce to 0.1 atm (state B). It takes 5 hours to carry out the change from state A to state B. The sample is then heated and the pressure is raised to 1 atm. In 20 sec the water is at 25°C (a return to state A). The internal energy change in going from state B to B is equal to, but opposite in sign from, the change from B to A.

_____

2. The work done in changing from A to B and that done in changing from B to A in statement 1 of this section are numerically the same but opposite in sign.

_____

3. At constant pressure the amount of heat absorbed or evolved by a system is called the enthalpy change, $\Delta H$.

_____

4. If volume does not change, the amount of heat released during a change of state of a system is equal to the decrease in internal energy of that system.

_____

5. If $\Delta S$ for a change in a system is positive, the change is said to be thermodynamically spontaneous.

_____    6. The more positive the value of $\Delta S_{total}$ for a change, the more rapid the change.

_____    7. For a spontaneous change the Gibbs free energy change is less than zero.

_____    8. As the reversible emf of a cell reaction becomes more negative, $\Delta G°$ becomes more positive.

_____    9. The standard free energy of formation of any element in its standard state is zero.

_____   10. The value of $\Delta G$ of a reaction is independent of pressure.

_____   11. The reaction of acetylene, H—C≡C—H, with oxygen is spontaneous.

_____   12. The value of Gibbs free energy of formation is positive for every compound.

_____   13. The entropy of a mole of solid sodium is less than that of a mole of gaseous sodium.

_____   14. More heat is released to the surroundings when a mole of $H_2$ is burned at constant pressure (an open flame) than when it is burned at constant volume (a bomb calorimeter).

II. Complete each of the following statements with an entry from the list on the right. An entry may be used more than once.

1. The energy of the universe is con-    (a) enthalpy
   stant, but the _____ tends
   toward a maximum.

2. The _____ of a reaction can be    (b) entropy
   measured indirectly with a bomb
   calorimeter.

3. The effective concentration of a    (c) zero
   substance is called the _____.

4. The amount of work done by a process
   carried out at constant volume is    (d) negative
   _____.

5. By convention, in the equation describing the change in internal
   energy of a system, $\Delta E = q - w$,    (e) positive
   work done by the system is always
   _____.

                                         (f) gram

6. Entropy, enthalpy, and internal      (g) atmosphere
   energy are _____ functions.

7. For any thermodynamically spontaneous change the total change in      (h) newton
   entropy must be a _____ value.

8. For any thermodynamically spontaneous change the Gibbs free energy    (i) activity
   must be a _____ value.

9. For a system at equilibrium $\Delta G°$
   equals _____.                      (j) state

10. The thermodynamic function that does
    not necessitate a consideration of
    the changes in surroundings is       (k) Gibbs free
    _____.                                energy

11. The $\Delta S$ of dissolution of NaCl in
    $H_2O$ is a _____ value.

III. Answer each of the following:

1. Use the information in Table 5.1 of your text and
   the heat of combustion of $CH_4$ at 25°C and 1 atm,
   $\Delta H = 803.3$ kJ/mol, to calculate $\Delta H$ for the following
   reaction at 25°C and 1 atm:

   $$CO_2(g) + 2H_2(g) \rightarrow 2H_2O(g) + C(s)$$

2. What is $\Delta E$ for the reaction described in problem
   1 above?

3. An unknown reaction is carried out in a bomb calorimeter containing only carbon and water. The heat of
   reaction at 25°C is found to be 131.4 kJ/mol. Which
   of the following reactions occurred?

   $$C(s) + 2H_2O(g) \rightarrow CO_2(g) + 2H_2(g)$$
   $$C(s) + H_2O(g) \rightarrow CO(g) + H_2(g)$$

4. Use the data in Table 14.1 of your text to determine
   the value of $\Delta H$ for each reaction in problem 3 above.
   The heat of atomization of C(s) is 720 kJ/mol and the
   $C\equiv O$ bond energy is 1080 kJ/mol. Compare values calculated from average bond energies with those calculated from values of $\Delta H_f°$ (see problem 3 above) and
   offer an explanation of any differences observed.

5. Determine the $C\!=\!\!=\!\!O$ bond energy in $CO_2$ from values
   given in your text and the study guide.

6. Use values of $\Delta G_f^o$ to determine $\Delta G$ for the oxidation of aluminum at 25°C and 1 atm:

$$4Al(s) + 3O_2(g) \rightarrow 2Al_2O_3(s)$$

7. Calculate $\mathscr{E}°$ for the reaction

$$2Al(s) = 3/2\ O_2(g) \rightarrow Al_2O_3(s)$$

8. Calculate $K_p$ for the reaction

$$C(s) + H_2O(g) \rightarrow CO(g) + H_2(g)$$

9. Use the data in Table 14.1 of your text to determine the standard enthalpy of formation of $CF_4(g)$. The heat of atomization of $C(s)$ is 720 kJ/mol.

10. Determine the average C—H bond energy from standard enthalpy of formation data in Table 5.1 of your text. The heat of atomization of $C(s)$ is 720 kJ/mol.

11. Use the results of the calculation in problem 10 above and standard enthalpy of formation data in Table 5.1 of your text to predict the C—C bond in ethane:

    Compare the calculated value with the C—C bond energy given in Table 14.1 of your text.

12. Use the values of $\Delta H_f^o$ and $\Delta G_f^o$ given in Tables 5.1 and 14.1 of your text to determine the change in entropy for the following reaction at 25°C and 1 atm:

$$CH_4(g) + 2O_2(g) \rightarrow CO_2(g) + 2H_2O(g)$$

13. Compare the value of $\Delta S$ calculated in problem 12 above with the value computed from the data in Table 14.5 of your text.

14. Calculate the equilibrium constant for the reaction

$$H_2O \rightarrow H^+(aq) + OH^-(aq)$$

    from the following electrode potentials:

$$2H_2O + 2e^- \rightarrow H_2(g) + 2OH^-(aq) \qquad \mathscr{E}° = -0.828\ V$$
$$2H^+(aq) + 2e^- \rightarrow H_2(g) \qquad \mathscr{E}° = 0.000\ V$$

15. When 1.46 g of adipic acid, $C_6H_{10}O_4$, is burned at 25°C and constant pressure, the products are $CO_2(g)$ and $H_2O(l)$. If 6.690 kcal (or 27.991 kJ) are released in the reaction, what is $\Delta H$ for the reaction of 1 mole of adipic acid?

## ANSWERS TO EXERCISES

I. Concepts of chemical thermodynamics

1. True
   [14.1]

Internal energy is a state function. The difference in the internal energies of the two states is inde!endent of the path taken between states.

2. False
   [14.1]

Work depends on the path, i.e., the method of going from one state to another.

3. True
   [14.2]

The relationship between $\Delta H$ and $\Delta E$ is

$$\Delta H = \Delta E + P\Delta V$$

At constant volume the amount of heat absorbed or evolved by a system is the internal energy change, $\Delta E$.

4. True
   [14.2]

At constant volume

$$\Delta H = \Delta E$$

and

$$q_V = q_p$$

5. False
   [14.4]

For a spontaneous change $\Delta S_{total}$ is positive. Since $\Delta S_{total} = \Delta S_{system} + \Delta S_{surroundings}$, it is possible to have a negative value of $\Delta S_{system}$ and still have a thermodynamically spontaneous reaction.

6. False
   [14.4]

A positive value of $\Delta S_{total}$ indicates that the change is thermodynamically favorable. The change, however, may not occur at an observable rate.

7. True
   [14.5]

8. False
   [14.9]

The relationship between $\Delta G°$ and $\mathcal{E}°$ is

$$\Delta G° = -nF\mathcal{E}°$$

The value of $\Delta G°$ becomes more positive as $\mathcal{E}°$ becomes more negative.

9. True
   [14.7]

10. False
    [14.6]

The values of $\Delta G$ and $\Delta S$ and the amount of work done by the system are not independent of pressure.

11. True
    [14.7]

The reaction is exothermic, as evidenced by the oxy-acetylene flames used in welding.

12. False
    [14.7]

Values of $\Delta G_f$ can be positive or negative.

13. True
    [14.4]

The solid is more ordered than the gas.

14. True At constant volume the heat released is
[14.2]

$$\Delta E = q_v$$

No pressure-volume work is done. At constant pressure

$$q_p = \Delta E + P\Delta V$$

Since the $\Delta V$ for the reaction is negative, the heat released to the surroundings by the burning of $H_2$ is more
at constant pressure than at constant volume:

$$H_2(g) + 1/2\ O_2(g) \rightarrow H_2O(g)$$

## II. Conventions and terms of thermodynamics

1. entropy [14.4]
2. enthalpy [14.2]
3. activity [14.9]
4. zero [14.2]
5. positive [14.1]
6. state [14.4]
7. positive [14.4]
8. negative [14.5]
9. zero [14.5]
10. Gibbs free energy
    [14.5]
11. positive [14.4]

## III. Chemical thermodynamic calculations

1. $\Delta H = -89.1$ kJ   We add the appropriate equations and the corresponding
   [14.2]                enthalpy values:

| | |
|---|---|
| $CH_4(g) \rightarrow C(s) + 2H_2(g)$ | $\Delta H = 74.85$ kJ |
| $CO_2(g) + 2H_2O(g) \rightarrow 2O_2(g) + CH_4(g)$ | $\Delta H = 803.3$ kJ |
| $4H_2(g) + 2O_2(g) \rightarrow 4H_2O(g)$ | $\Delta H = 4(-241.8)$ kJ |
| $CO_2(g) + 2H_2(g) \rightarrow 2H_2O + C(s)$ | $\Delta H = -89.05$ kJ |

2. $\Delta E = -86.6$ kJ   The relationship between $\Delta E$ and $\Delta H$ is
   [14.2]

$$\Delta E = \Delta H - \Delta nRT$$

Substituting values into the preceding equation, we find

$$\Delta E = \Delta H - \Delta nRT$$
$$= 89.05\ \text{kJ} - (-1)(8.314 \times 10^{-3}\ \text{kJ/K·mol})(298\ \text{K})$$
$$= -86.57\ \text{kJ}$$

3. $C(s) + 2H_2O(g)$   We use the information in Table 14.1 of your text to cal-
   $\rightarrow CO_2(g) + 2H_2(g)$   culate the theoretical value of $\Delta H$ for each reaction and
   the information in the problem to calculate the experi-
   mental value of $\Delta H$ for each reaction. For the reaction:

$$C(s) + 2H_2O(g) \rightarrow CO_2(g) + 2H_2(g)$$

we first determine the value of $\Delta H$ from values of $\Delta H_f^\circ$:

$$\Delta H = \Delta H_f^\circ(CO_2) - 2\Delta H_f^\circ(H_2O)$$

$$= -393.5 \text{ kJ} - 2(-241.8 \text{ kJ})$$

$$= 90.1 \text{ kJ}$$

For the reaction

$$C(s) + H_2O(g) \rightarrow CO(g) + H_2(g)$$

we determine the value of $\Delta H$ from values of $\Delta H_f^\circ$:

$$\Delta H = \Delta H_f^\circ(CO) - \Delta H_f^\circ(H_2O)$$

$$= -110.5 \text{ kJ} - (-241.8 \text{ kJ})$$

$$= 131.3 \text{ kJ}$$

We then determine the value of $\Delta H$ from experimental data:

$$\Delta H = \Delta E + \Delta nRT$$

$$= 131.3 \text{ kJ} + (1)(8.314 \times 10^{-3} \text{ kJ/K·mol})(298 \text{ K})$$

$$= 131.6 \text{ kJ}$$

The values of $\Delta H$ determined from values of $\Delta H_f^\circ$ and from experimental data for the reaction

$$C(s) + 2H_2O(g) \rightarrow CO_2(g) + 2H_2(g)$$

are in closer agreement than those values of $\Delta H$ determined for the reaction

$$C(s) + H_2O(g) \rightarrow CO(g) + H_2(g)$$

Therefore, the water and carbon react in the calorimeter to form carbon dioxide and hydrogen.

4. 293 kJ
   131 kJ
   [14.3]

We add the appropriate equations and the corresponding enthalpy values. For the reaction

$$C(s) + 2H_2O(g) \rightarrow CO_2(g) + 2H_2(g)$$

we find

| | |
|---|---|
| $2H_2O(g) \rightarrow 4H(g) + 2O(g)$ | $\Delta H = 4(463)$ kJ |
| $C(s) \rightarrow C(g)$ | $\Delta H = 720$ kJ |
| $C(g) + 2O(g) \rightarrow O{=\!=}C{=\!=}O(g)$ | $\Delta H = 2(-707)$ kJ |
| $4H(g) \rightarrow 2H_2(g)$ | $\Delta H = 2(-435)$ kJ |
| $C(s) + 2H_2O(g) \rightarrow CO_2(g) + 2H_2(g)$ | $\Delta H = 293$ kJ |

Note that four O—H bonds are broken, and two C$=$O and two H—H bonds are formed.

The value of $\Delta H$ determined from average bond energies is higher by 45 kcal than that determined from values of $\Delta H_f^\circ$. The discrepancy probably occurs because the breaking of the second C$=$O bond in $CO_2$ requires more energy than the breaking of the first. Thus, the average bond energy value of C=O in Table 14.1 of your text is not the average bond energy of the C=O bonds in $CO_2$.

For the reaction

$$C(s) + H_2O(g) \rightarrow CO(g) + H_2(g)$$

we find

| | |
|---|---|
| $H_2O(g) \rightarrow 2H(g) + O(g)$ | $\Delta H = 2(463)$ kJ |
| $C(s) \rightarrow C(g)$ | $\Delta H = 720$ kJ |
| $C(g) + O(g) \rightarrow CO(g)$ | $\Delta H = -1080$ kJ |
| $2H(g) \rightarrow H_2(g)$ | $\Delta H = 435$ kJ |
| $C(s) + H_2O(g) \rightarrow CO(g) + H_2(g)$ | $\Delta H = 131$ kJ |

The value of $\Delta H$ calculated from average bond energies is higher than that calculated from values of $\Delta H_f^\circ$.

5. 804 kJ/mol
[14.3]

The enthalpy of formation of a C=O bond in $CO_2$ is one-half the enthalpy of formation of 1 mole of $CO_2(g)$, i.e., one-half the enthalpy of the reaction:

$$C(g) + 2O(g) \rightarrow CO_2(g)$$

We can determine $\Delta H_f$ of $CO_2(g)$ as follows:

| | |
|---|---|
| $C(s) + O_2(g) \rightarrow CO_2(g)$ | $\Delta H^\circ = -393.5$ kJ |
| $C(g) \rightarrow C(s)$ | $\Delta H_{fusion} = -720$ kJ |
| $2O(g) \rightarrow O_2(g)$ | $\Delta H = -494$ kJ |
| $C(g) + 2O(g) \rightarrow CO_2(g)$ | $\Delta H = -1607.5$ kJ |

Since there are two C=O bonds in $CO_2$, the average bond energy of C=O in $CO_2$ is -804 kJ:

$$\frac{-1607.5 \text{ kJ}}{2} = -803.7 \text{ kJ}$$

Note the difference in this value and the one given in Table 14.1 of your text.

6. $\Delta G = -3152.8$ kJ
[14.7]

The equation

$$4Al(s) + 3O_2(g) \rightarrow 2Al_2O_3(s)$$

represents the formation of 2 mol of $Al_2O_3$ at 25°C and 1 atm. From Table 14.5 of your text we note

$$2Al(s) + 3/2\ O_2(g) \rightarrow Al_2O_3(s) \qquad \Delta G_f^\circ = -1576.4 \text{ kJ}$$

Thus,

$$4Al(s) + 3O_2(g) \rightarrow 2Al_2O_3(s) \qquad \Delta G_f^\circ = -3152.8 \text{ kJ}$$

7. 2.7230 V

The relationship between $\Delta G^\circ$ and $\mathscr{E}^\circ$ is $\Delta G^\circ = -nF\mathscr{E}^\circ$. Rearranging the preceding equation and substituting values into it, we find

$$\mathscr{E}° = -\frac{\Delta G°}{nF}$$

$$= -\frac{(-3152.8 \times 10^6 \text{ J})}{(12)(96500 \text{ coul})}$$

$$= 2.7230 \text{ V}$$

8. $K_p = 1.01 \times 10^{-16}$   Since
   [14.9]

$$\Delta G° = -2.303RT \log K_p$$

we must determine $\Delta G°$ for the reaction:

$$\Delta G° = \Delta G_f°(CO) - \Delta G_f°(H_2O)$$

$$= -137.28 \text{ kJ} - (-228.61 \text{ kJ})$$

$$= 91.33 \text{ kJ}$$

Rearranging

$$\Delta G° = -2.303RT \log K_p$$

and substituting values into it, we find

$$\log K_p = -\frac{\Delta G°}{2.303RT}$$

$$= \frac{91.33 \text{ kJ/mol}}{(2.303)(8.314 \times 10^{-3} \text{ kJ/K·mol})(298.2 \text{ K})}$$

$$= -15.996$$

and

$$K_p = 1.01 \times 10^{-16}$$

9. −910 kJ     We add the appropriate equations and corresponding values
   [14.3]      of

| | |
|---|---|
| $C(s) \to C(g)$ | $\Delta H = 720 \text{ kJ}$ |
| $2F_2(g) \to 4F(g)$ | $\Delta H = 2(155 \text{ kJ})$ |
| $C(g) + 4F(g) \to CF_4(g)$ | $\Delta H = 4(-485 \text{ kJ})$ |
| $C(s) + 2F_2(g) \to CF_4(g)$ | $\Delta H = -910 \text{ kJ}$ |

The value of $\Delta H_f°$ for $CF_4(g)$ from Table 5.1 of your
text is −913.4 kJ.

10. 416 kJ     Adding equations:
    [14.3]

| | |
|---|---|
| $C(s) + 2H_2(g) \to CH_4(g)$ | $\Delta H = -74.85 \text{ kJ}$ |
| $C(g) \to C(s)$ | $\Delta H = -720 \text{ kJ}$ |
| $4H(g) \to 2H_2(g)$ | $\Delta H = 2(-435 \text{ kJ})$ |
| $C(g) + 4H(g) \to CH_4(g)$ | $\Delta H = -1664.85 \text{ kJ}$ |

The enthalpy for forming all bonds in $CH_4$ is 1665 kJ/mol.
This is the enthalpy of forming four C—H bonds. Thus the
enthalpy for forming a single C—H bond is −416 kJ/mol.
Therefore the enthalpy required to break such a bond is
+416 kJ/mol.

11. -346 kJ
   [14.6]

Adding equations:

$$2C(s) + 3H_2(g) \rightarrow C_2H_6(g) \qquad \Delta H = -84.68 \text{ kJ}$$
$$2C(g) \rightarrow 2C(s) \qquad \Delta H = 2(-720 \text{ kJ})$$
$$6H(g) \rightarrow 3H_2(g) \qquad \Delta H = 3(-435 \text{ kJ})$$

overline

$$2C(g) + 6H(g) \rightarrow C_2H_6(g) \qquad \Delta H = -2829.68 \text{ kJ}$$

Since -2830 kJ/mol is $\Delta H$ for forming all bonds, we sub-tract the enthalpy for forming six C—H bonds to obtain the enthalpy for forming the single C—C bond:

$$(C\text{—}C \text{ bond}) = -2830 \text{ kJ} - 6(-414 \text{ kJ})$$

$$= -346 \text{ kJ}$$

12. -4.83 J/K
   [14.5]

First determine $\Delta H$ for the reaction:

$$\Delta H = \Delta H_f^\circ(CO_2) + 2\Delta H_f^\circ(H_2O) - \Delta H_f^\circ(CH_4)$$

$$= -393.5 \text{ kJ} + 2(-241.8 \text{ kJ}) - (-74.85 \text{ kJ})$$

$$= -802.25 \text{ kJ}$$

Then determine $\Delta G$ for the reaction:

$$\Delta G = \Delta G_f^\circ(CO_2) + 2\Delta G_f^\circ(H_2O) - \Delta G_f(CH_4)$$

$$= -394.38 + 2(-228.61 \text{ kJ}) - (-50.79 \text{ kJ})$$

$$= -800.81 \text{ kJ}$$

Finally calculate $\Delta S$ for the reaction:

$$\Delta S = \frac{\Delta H - \Delta G}{T}$$

$$= \frac{(-802.25 \text{ kJ}) - (-800.81 \text{ kJ})}{298.2 \text{ K}}$$

$$= -4.83 \text{ J/K}$$

13. -5.3 J/K
   [14.8]

First calculate $\Delta S^\circ$:

$$\Delta S = [S^\circ(CO_2) + 2S^\circ(H_2O)] - [S^\circ(CH_4) + 2S^\circ(O_2)]$$

$$= 213.6 \text{ J/K} + 2(188.7 \text{ J/K}) - 186.2 \text{ J/K} - 2(205.03 \text{ J/}$$

$$= -5.26 \text{ J/K}$$

14. $K_p = 1.02 \times 10^{-14}$
   [14.7, 14.9]

Add the following half reactions and corresponding values of $\mathcal{E}^\circ$ to determine the value of $\mathcal{E}^\circ$ for the ionization of water:

$$2H_2O + 2e^- \rightarrow H_2 + 2OH^- \qquad \mathcal{E}^\circ = -0.828 \text{ V}$$
$$H_2 \rightarrow 2H^+ + 2e^- \qquad \mathcal{E}^\circ = 0.000 \text{ V}$$

overline

$$2H_2O \rightarrow 2H^+ + 2OH^- \qquad \mathcal{E}^\circ = -0.828 \text{ V}$$

Since

$$\Delta G^\circ = -nF\mathcal{E}^\circ$$

and

$$\Delta G° = -2.303RT \log K_p$$

then

$$-nF\mathcal{E}° = -2.303RT \log K_p$$

and

$$\log K_p = \frac{nF\mathcal{E}°}{2.303}$$

Substituting values into the preceding equation, we find

$$\log K_p = \frac{nF\mathcal{E}°}{2.303RT}$$

$$= -13.99$$

$$= 1.02 \times 10^{-14}$$

15. -2799 kJ
    [14.2]

First write a balanced equation for the reaction:

$$2C_6H_{10}O_4(s) + 13O_2(g) \rightarrow 10H_2O(l) + 12CO_2(g)$$

Then compute the number of moles of $C_6H_{10}O_4$ in 1.46 g of $C_6H_{10}O_4$:

$$? \text{ mol } C_6H_{10}O_4 = 1.46 \text{ g } C_6H_{10}O_4\left(\frac{1 \text{ mol } C_6H_{10}O_4}{146.1 \text{ g } C_6H_{10}O_4}\right)$$

$$= 1.00 \times 10^{-2} \text{ mol } C_6H_{10}O_4$$

We know that 27.991 kJ is released when $1.00 \times 10^{-2}$ mol of $C_6H_{10}O_4$ is burned. Therefore, we can calculate the amount of heat released by 1 mol of $C_6H_{10}O_4$:

$$\Delta H = \left(\frac{27.991 \text{ kJ}}{0.01 \text{ mol } C_6H_{10}O_4}\right)(1 \text{ mol } C_6H_{10}O_4)$$

$$= 2799 \text{ kJ}$$

## SELF-TEST

Complete the test in 35 minutes:

1. The heat of combustion of oxalic acid, $H_2C_2O_4$, is -253 kJ/mol at 25°C and constant volume. What is $\Delta H$?

2. Using standard electrode potentials at 25°C, calculate $K_p$ for

   $$Cu^{2+}(aq) + Sn(s) \rightarrow Cu(s) + Sn^{2+}(aq)$$

3. Using standard entropies and enthalpies of formation determine if the reaction

   $$CO(g) + H_2O(g) \rightarrow CO_2(g) + H_2(g)$$

   is spontaneous as written and calculate $K_p$.

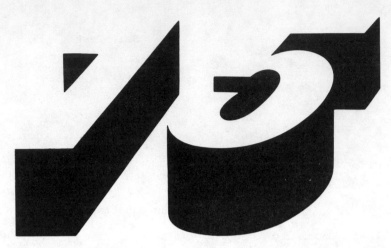

# Acids and Bases

OBJECTIVES

I. You should be able to demonstrate your knowledge of the following terms by defining them, describing them, or giving specific examples of them:

acid [15.1, 15.2, 15.3]
amphiprotic [15.3]
anhydride [15.1]
Arrhenius concept [15.1]
base [15.1, 15.2, 15.3]
Brønsted-Lowry concept [15.3]
conjugate pair [15.3]
electrophilic [15.7]
hydrolysis [15.5]
leveling effect [15.4]
Lewis concept [15.7]
neutralization [15.1, 15.2]
nucleophilic [15.7]
solvent system concept [15.2]

232

II. You should be able to identify acids and bases in given reactions and deduce relative strengths of each from displacement of equilibrium and/or from a qualitative knowledge of factors that influence the strength of each.

**EXERCISES**    I. Fill in each of the statements with the appropriate entry or entries from the following list. An entry may be used more than once.

(a) ammonium ion
(b) chloride ion
(c) hydrofluoric acid
(d) hydroiodic acid

(e) sodium acetate
(f) water
(g) zinc(II) ion

1. In glacial acetic acid, i.e., pure acetic acid, _____ would be a strong base.

2. The strongest acid that can exist in liquid ammonia is _____.

3. _____ is the weakest hydro acid of the elements of group VII A.

4. _____ are species that can act as Brønsted acids in water.

5. _____ are species that can act as Brønsted bases in water.

6. _____ is a species that can act as a Lewis acid.

7. _____ are species that can act as Lewis bases.

8. The strongest electrophilic species is _____.

9. The conjugate acid of ammonia is _____.

10. _____ is an amphiprotic material.

11. If 0.1 mol of _____ were added to 1 liter of water, a basic solution would be formed. Assume that the species are added as chloride salts of cations and sodium salts of anions.

12. If 0.1 mol of _____ were added to 1 liter of water, a neutral solution would be formed. Assume that the species are added as chloride salts of cations and sodium salts of anions.

II. Answer each of the following. Assume that each equilibrium is displaced to the right.

1. The strongest Lewis acid in the reaction

$$HCl + H_2O \rightleftharpoons Cl^- + H_3O^+$$

is _____.

2. The strongest Brønsted base in the reaction

$$H_2SO_4 + F^- \rightleftharpoons HSO_4^- + HF$$

is _____.

3. The strongest Brønsted acid in the reaction

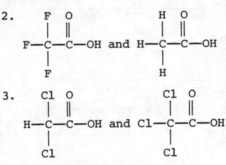

is _____.

III. Use qualitative knowledge of factors that influence the strength of an acid to answer the following:

For each pair of compounds, determine which compound is the stronger acid:

1. HI and HF

2.
$$\underset{\underset{F}{|}}{\overset{F}{|}}{F-C-}\overset{O}{\overset{||}{C}}-OH \quad \text{and} \quad \underset{\underset{H}{|}}{\overset{H}{|}}{H-C-}\overset{O}{\overset{||}{C}}-OH$$

3.
$$\underset{\underset{Cl}{|}}{\overset{Cl}{|}}{H-C-}\overset{O}{\overset{||}{C}}-OH \quad \text{and} \quad \underset{\underset{Cl}{|}}{\overset{Cl}{|}}{Cl-C-}\overset{O}{\overset{||}{C}}-OH$$

4. $HClO_4$ and $HClO_3$

5. HOI and HOCl

IV. Use qualitative knowledge of factors that influence the strength of an acid or a base to answer the following. Assume that each equilibrium is displaced to the right.

1. List all Brønsted acids in the following reactions in order of decreasing strength:
(a) $HOBr + CN^- \rightleftharpoons OBr^- + HCN$
(b) $HCl + H_2O \rightleftharpoons H_3O^+ + Cl^-$
(c) $HI + BrO_3^- \rightleftharpoons HBrO_3 + I^-$
(d) $OH^- + HOBr \rightleftharpoons H_2O + OBr^-$
(e) $H_3O^+ + IO_3^- \rightleftharpoons HIO_3 + H_2O$

2. Which acids in problem 1 above are strong acids?

3. List the Brønsted bases in problem 1 above in order of decreasing strength.

4. Identify the Lewis acid in the following reaction:

$$PCl_3 + Cl_2 \rightarrow PCl_5$$

## ANSWERS TO EXERCISES

I. Acid—base systems

1. sodium acetate [15.4]

The acetate ion is the strongest base that can exist in acetic acid. The basicity of strongly basic materials in acetic acid is leveled to that of the acetate ion.

2. ammonium ion [15.4]

3. hydrofluoric acid [15.4]

There is an increase in acid strength of the hydro acids, HX, of the elements in any group of the periodic table with increasing atomic size of the electronegative element, X.

4. ammonium ion hydrofluoric acid hydroiodic acid water [15.3]

A Brønsted acid is a species that can donate a proton:

| $NH_4^+$ | + | $OH^-$ | $\rightleftharpoons$ | $NH_3$ | $H_2O$ |
|---|---|---|---|---|---|
| Brønsted acid | | Brønsted base | | Brønsted base | Brønsted acid |

| HF | + | $H_2O$ | $\rightleftharpoons$ | $F^-$ | $H_3O^+$ |
|---|---|---|---|---|---|
| Brønsted acid | | Brønsted base | | Brønsted base | Brønsted acid |

| HI | + | $H_2O$ | $\rightleftharpoons$ | $I^-$ | $H_3O^+$ |
|---|---|---|---|---|---|
| Brønsted acid | | Brønsted base | | Brønsted base | Brønsted acid |

| $H_2O$ | + | $H^-$ | $\rightleftharpoons$ | $H_2$ | $OH^-$ |
|---|---|---|---|---|---|
| Brønsted acid | | Brønsted base | | Brønsted base | Brønsted acid |

5. chloride ion sodium acetate water [15.3]

A Brønsted base is a species that can accept a proton:

| $Cl^-$ | + | HI | $\rightleftharpoons$ | HCl | $I^-$ |
|---|---|---|---|---|---|
| Brønsted base | | Brønsted acid | | Brønsted acid | Brønsted base |

| $C_2H_5O_2^-$ | + | HCl | $\rightleftharpoons$ | $C_2H_5O_2H$ | + | $Cl^-$ |
|---|---|---|---|---|---|---|
| Brønsted base | | Brønsted acid | | Brønsted acid | | Brønsted base |

| $H_2O$ | + | HBr | $\rightleftharpoons$ | $H_3O^+$ | $Br^-$ |
|---|---|---|---|---|---|
| Brønsted base | | Brønsted acid | | Brønsted acid | Brønsted base |

6. Zn(II) ion
   [15.7]

A Lewis acid *accepts* an electron pair to form a covalent bond.

7. chloride ion
   [15.7]

A Lewis base *donates* an electron pair to form a covalent bond.

8. Zn(II) ion
   [15.7]

The word *electrophilic* is used to describe the ability of a species to attract an electron pair.

9. ammonium ion

$$H_2O + NH_3 \rightleftharpoons NH_4^+ + OH^-$$
$$\text{acid}_1 \quad \text{base}_1 \quad \text{acid}_2 \quad \text{base}_1$$

10. water
    [15.3]

Water can function as an acid and as a base:

$$H_2O + NH_2^- \rightleftharpoons NH_3 + OH^-$$

$$H_2O + HBr \rightleftharpoons Br^- + H_3O^+$$

11. sodium acetate
    [15.5]

Acetate ion hydrolyzes:

$$H_2O + C_2H_3O_2^- \rightleftharpoons HC_2H_3O_2 + OH^-$$

12. chloride ion
    water
    [15.1]

II. Strengths of acids and bases determined from equilibrium displacement

1. $H^+$
   [15.7]

The hydrogen ion, $H^+$, from HCl is the electron acceptor.

2. $F^-$
   [15.3]

The fluoride ion, $F^-$, is the proton acceptor.

3. HF
   [15.3]

Hydrogen fluoride, HF, is the proton donor.

III. Strengths of acids and bases determined from qualitative knowledge of factors that influence strength

1. HI
   [15.6]

2. 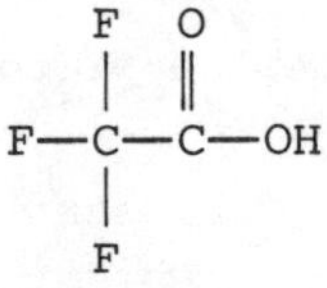
```
      F   O
      |   ||
  F—C—C—OH
      |
      F
```
   [15.6]

3. 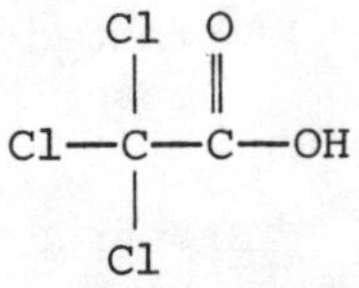
```
      Cl  O
      |   ||
  Cl—C—C—OH
      |
      Cl
```
   [15.6]

4. $HClO_4$
[15.6]

5. HOCl
[15.6]

IV. Relative strengths of acids and bases determined from information on equilibrium displacement and qualitative knowledge of factors that influence strength

1. HI
HCl

The order of acid strength of the hydrogen halides is HI > HBr > HCl > HF.

$H_3O^+$

Hydrochloric acid, HCl, is stronger than the hydronium ion, $H_3O^+$ (see reaction c).

$HBrO_3$
$HIO_3$

Iodic acid, $HIO_3$, is weaker than the hydronium ion, $H_3O^+$ (see reaction f). Bromic acid, $HBrO_3$, is expected to be more acidic than iodic acid, $HIO_3$, and less acidic than hydroiodic acid, HI (see reaction d).

HOBr

Hydrobromous acid, HOBr, is stronger than HCN (see reaction a) and $H_2O$ (see reaction e).

HCN
$H_2O$
[15.6]

2. HCl and HI

Hydrochloric acid and hydrobromic acid are leveled by water:

$$HCl + H_2O \rightleftharpoons Cl^- + H_3O^+$$
$$HI + H_2O \rightleftharpoons I^- + H_3O^+$$

3. $OH^-$
$CN^-$
$OBr^-$
$IO_2^-$
$IO_3^-$
$BrO_3^-$
$H_2O$
$Cl^-$
$I^-$
[15.6]

Increasing acid strength parallels decreasing base strength of the conjugate base. Thus, the conjugate base of the strongest acid is the weakest base. Bases weaker than water are not normally considered to be basic species in aqueous solution.

4. $PCl_3$
[15.7]

Trichlorophosphine, $PCl_3$, is a Lewis acid and accepts an electron pair from each chlorine atom of $Cl_2$ to form $PCl_5$.

SELF-TEST

_____

_____

_____

_____

_____

_____

_____

Complete the test in 10 minutes:

1. The reaction of ammonia and the hydride ion goes to completion:

$$NH_3 + H^+ \rightarrow H_2 + NH_2^-$$

What is the strongest Brønsted acid in the reaction? What is the strongest base?

2. What is the conjugate acid of $HS^-$?

3. The following reaction may be interpreted as an acid–base reaction in a $SO_2$ solvent system:

$$Cs_2SO_3 + SOCl_2 \rightarrow 2CsCl + 2SO_2$$

What is the acid?

4. What is the Lewis base in the following reaction?

$$BF_3 + F^- \rightarrow BF_4^-$$

5. What is the Lewis acid in the following reaction?

$$[:\overset{..}{S}:]^{2-} + \overset{..}{S}: \rightarrow S_2^{2-}$$

6. The following equilibrium is displaced to the right:

$$HC_2H_3O_2 + HS^- \rightleftharpoons H_2S + C_2H_3O_2^-$$

What is the strongest Brønsted base in the reaction? What is the strongest Brønsted acid?

7. What is the conjugate base of $H_2PO_4^-$?

# Ionic Equilibria, Part I

OBJECTIVES

I. You should be able to demonstrate your knowledge of the following terms by defining them, describing them, or giving specific examples of them:

buffer [16.2]
common-ion effect [16.6]
degree of dissociation, α [16.1]
glass electrode [16.4]
hydronium ion [16.1]
indicator [16.5]
$K_W$ [16.2]
pH [16.3, 16.4]
p$K$ [16.7]
polyprotic acids [16.8]
pOH [16.3]
strong electrolyte [16.1]
weak electrolyte [16.1]

II. You should be able to calculate the pH and the concentration of all species in a solution containing either strong acids or bases, weak acids or bases, buffers, or polyprotic acids.

III. You should be able to determine the ionization constant of a weak acid from potentiometric titration data or from concentrations of all species in solution.

IV. You should be able to determine the pH of a solution from cell potential data.

UNITS,
SYMBOLS,
MATHEMATICS

I. It is imperative that you be able to solve the quadratic equation. For an equation that contains the unknown to the second power as well as the unknown to the first power, rewrite it in the form:

$$ax^2 + bx + c = 0$$

Here the values of $a$, $b$, and $c$ are constants that can be either positive, negative, or zero, and $x$ represents the unknown. The values of $x$, which are proper solutions of the equation, are found by solving

$$x = \frac{-\phantom{b} \pm \sqrt{b^2 - 4ac}}{2a}$$

See example 16.2 of your text. You should review the use of logarithms in the appendix of your text and Chapter 14 of the study guide.

Remember that pH values have no units.

EXERCISES

I. Do the following:

1. The parietal cells of the human stomach secrete 0.155$M$ HCl. What is the pH of the secretion?

2. What is the pH of a solution with a [OH$^-$] of $4.8 \times 10^{-4}M$?

3. A certain enzyme loses its ability to catalyze a reaction below a pH of 5.70. To what [H$^+$] does the pH correspond?

4. What are the pH and pOH of a 0.020$M$ solution of KOH?

5. Formic acid, HCOOH, has an ionization constant of $1.77 \times 10^{-4}$. What is the pH of 0.10$M$ formic acid?

6. What are the concentrations of $H^+$, $ClO^-$, and $HClO$ in 0.10$M$ hypochlorous acid? The ionization constant of hypochlorous acid is $3.2 \times 10^{-8}$.

7. How many moles of nitrous acid, $HNO_2$, must be added to water to prepare 1.0 liter of solution with a pH of 3.0? The ionization constant of nitrous acid is $4.5 \times 10^{-4}$.

8. What is the concentration of a solution in which HX is 1.0 percent dissociated? The ionization constant of HX is $5.0 \times 10^{-4}$.

9. In metabolic acidosis the $HCO_3^-/CO_2$ ratio in the blood may be 16/1, a ratio that corresponds to a pH of 7.3. What is the p$K$ of carbonic acid?

10. In metabolic alkalosis the $HCO_3^-/CO_2$ ratio in the blood may be 40/1. What is the pH of blood under such conditions? Use the value of the ionization constant of carbonic acid calculated in problem 9 above.

11. A cell consisting of a hydrogen electrode and a normal calomel elextrode contains a solution of unknown pH. The potential is 0.833 V at 25°C. Calculate the following:
    (a) the pH of the oslution
    (b) the $[H^+]$ of the solution

12. The indicator bromothymol blue has an ionization constant of $5.0 \times 10^{-8}$. The acid color is yellow, and the alkaline color is blue. The acid color is visible when the ratio of yellow to blue is 20 to 1, and the blue is visible when the ratio of blue to yellow is 2 to 1. What is the pH range of color change of bromothymol blue?

13. A solution is prepared by adding 0.040 mol of solid NaCNO, sodium cyanate, to 250 ml of 0.025$M$ cyanic acid, HCNO. Assume that no volume change occurs. Calculate the pH of the solution. The ionization constant of cyanic acid is $1.2 \times 10^{-4}$.

14. Strychnine is a weak base, and its aqueous ionization is represented by

    $$S(aq) + H_2O \rightleftharpoons SH^+(aq) + OH^-(aq)$$

    A 1.0$M$ solution of strychnine has a pH of 11.00. What is the ionization constant of strychnine?

15. What is the pH of a buffer that is 0.15$M$ in NaZ and 3.0$M$ in HZ if the ionization constant of the weak acid HZ is $5.0 \times 10^{-6}$?

16. How many moles of solid $NH_4Cl$ must be dissolved in 500.0 ml of a solution that is originally $0.250M$ in $NH_3$ in order to have a final pH of 10.0? Assume that the volume remains 500.0 ml. The ionization constant of $NH_3$ is $1.8 \times 10^{-5}$.

17. Calculate the pH of a solution composed of 50 ml of $1.0M$ $NaC_2H_3O_2$ and 50 ml of $1.0M$ $HC_2H_3O_2$. Assume the volume after mixing to be 100 ml. The ionization constant of acetic acid, $HC_2H_3O_2$, is $1.8 \times 10^{-5}$.

18. What is the $[H^+]$ in a $0.036M$ $H_2S$ solution? For $H_2S$ the ionization constant of the primary ionization is $1.1 \times 10^{-7}$. Assume that effects due to secondary ionization are negligible.

19. Vitamin C, ascorbic acid, is a diprotic acid with the formula $C_6H_8O_6$. Calculate the $[H^+]$, $[C_6H_7O_6^-]$, and $[C_6H_6O_6^{2-}]$ of a $0.10M$ solution of ascorbic acid. The ionization constants of ascorbic acid are $K_1 = 7.9 \times 10^{-5}$ and $K_2 = 1.6 \times 10^{-12}$.

20. A buffer solution is prepared by mixing 500 ml of $1.00M$ $HC_2H_3O_2$ and 500 ml of $1.00M$ $NaC_2H_3O_2$. The ionization constant of acetic acid, $HC_2H_3O_2$, is $1.8 \times 10^{-5}$. Calculate the following:
    (a) the pH of the buffer
    (b) the pH of the buffer solution after the addition of 0.005 mol of HCl

## ANSWERS TO EXERCISES

I. Ionic equilibrium calculations

1. pH = 0.810 [16.3]

The definition of pH is

$$pH = -\log [H^+]$$

Substituting the value of $[H^+]$ into the preceding equation, we find

$$\begin{aligned} pH &= -\log [H^+] \\ &= -\log (1.55 \times 10^{-1}) \\ &= (-\log 1.55) + (-\log 10^{-1}) \\ &= 0.810 \end{aligned}$$

Alternatively, use your calculator, but review the proper manipulation of logs in the instruction manual.

2. pH = 10.68 [16.3]

The water constant at 25°C is

$$[H^+][OH^-] = 1.0 \times 10^{-14}$$

If we solve the preceding equation for $[H^+]$ and substitute values into the equation, we find

$$[H^+] = \frac{1.0 \times 10^{-14}}{4.8 \times 10^{-4}}$$
$$= 2.08 \times 10^{-11}M$$

We determine pH:

$$\begin{aligned} pH &= \log\ [H^+] \\ &= -\log\ [2.08 \times 10^{-11}) \\ &= (-\log 2.08) + (-\log 10^{-11}) \\ &= -0.32 + 11 \\ &= 10.68 \end{aligned}$$

3. $[H^+] = 2.0 \times 10^{-6}M$
[16.3]

We can determine $[H^+]$ as follows:

$$\begin{aligned} -\log\ [H^+] &= pH \\ \log\ [H^+] &= -pH \\ &= -5.70 \\ &= 0.30 - 6.0 \\ [H^+] &= (\text{antilog } 0.30)(\text{antilog } -6.0) \\ &= 2.0 \times 10^{-6}M \end{aligned}$$

We can also use another method of determining $[H^+]$:

$$\begin{aligned} pH &= -\log\ [H^+] \\ [H^+] &= 10^{-pH} \\ &= 10^{-5.7} \\ &= (10^{+0.3})(10^{-6}) \\ &= 2.0 \times 10^{-6}M \end{aligned}$$

4. pH = 12.30
pOH = 1.70
[16.3]

Potassium hydroxide is a strong electrolyte. Thus,

$$[OH^-] = 2.0 \times 10^{-2}M$$

and

$$\begin{aligned} pOH &= -\log\ [OH^-] \\ &= -\log\ (2.0 \times 10^{-2}) \\ &= -0.30 + 2 \\ &= 1.70 \end{aligned}$$

We determine the pH as follows: The water constant is

$$[H^+][OH^-] = 1.0 \times 10^{-14}$$

We arrange the preceding equation and substitute values into it:

$$[H^+] = \frac{1.0 \times 10^{-14}}{[OH^-]}$$
$$= \frac{1.0 \times 10^{-14}}{2.0 \times 10^{-2}}$$
$$= 5.0 \times 10^{-13}M$$

We calculate pH in either of two ways:

$$pH = -\log [H^+]$$
$$= -\log (5.0 \times 10^{-13})$$
$$= -0.70 + 13$$
$$= 12.30$$

or

$$pH + pOH = 14.00$$
$$pH = 14.00 - pOH$$
$$= 14.00 - 1.7$$
$$= 12.30$$

5. pH = 2.38
[16.1, 16.4]

The expression for the ionization constant is

$$K = 1.77 \times 10^{-4} = \frac{[H^+][HCOO^-]}{[HCOOH]}$$

We make the following assumptions:

| Species | Equilibrium Concentration |
|---------|---------------------------|
| $H^+$ | $(x)M$ |
| $HCOO^-$ | $(x)M$ |
| $HCOOH$ | $((1.0 \times 10^{-1}) - x \cong 1.0 \times 10^{-1})M$ |

We substitute concentrations into the expression for the ionization constant:

$$K = 1.77 \times 10^{-4} = \frac{[H^+][HCOO^-]}{[HCOOH]}$$

$$1.77 \times 10^{-4} = \frac{x^2}{1.0 \times 10^{-1}}$$

$$x^2 = 1.77 \times 10^{-5}$$

$$= 4.21 \times 10^{-3}$$

Thus, $[H^+] = 4.21 \times 10^{-3}M$ and the pH is

$$pH = -\log [H^+]$$
$$= -\log (4.21 \times 10^{-3})$$
$$= 2.38$$

6. $[H^+] = 5.7 \times 10^{-5}M$
$[ClO^-] = 5.7 \times 10^{-5}M$
$[HClO] = 1.0 \times 10^{-1}M$
[16.1]

The expression for the ionization constant is

$$K = 3.2 \times 10^{-8} = \frac{[H^+][ClO^-]}{[HClO]}$$

We make the following assumptions:

| Species | Original Concentration | Equilibrium Concentration |
|---------|------------------------|---------------------------|
| $H^+$ | $10^{-7}M$ | $(10^{-7} + x \cong x)M$ |
| $ClO^-$ | $0$ | $(x)M$ |
| $HClO$ | $1.0 \times 10^{-1}M$ | $((1.0 \times 10^{-1}) - x \cong 1.0 \times 10^{-1})M$ |

We substitute concentrations into the expression for the ionization constant:

$$3.2 \times 10^{-8} = \frac{[H^+][ClO^-]}{[HClO]}$$

$$= \frac{x^2}{1.0 \times 10^{-1}}$$

$$x^2 = 3.2 \times 10^{-9}$$

$$x = 5.7 \times 10^{-5}$$

Thus,

$$[H^+] = 5.7 \times 10^{-5} M$$

7. $3.2 \times 10^{-3}$ mol
   $HNO_2$
   [16.1, 16.3]

The equation for the ionization of nitrous acid is

$$HNO_2 \rightleftharpoons H^+ + NO_2^-$$

A pH of 3.0 corresponds to a $[H^+]$ of $1.0 \times 10^{-3} M$:

$$[H^+] = 10^{-pH}$$
$$= 10^{-3.0}$$
$$= 1.0 \times 10^{-3} M$$

We make the following assumptions:

| Species | Original Concentration | Equilibrium Concentration |
|---------|------------------------|---------------------------|
| $H^+$ | $10^{-7} M$ | $(10^{-7} + (1.0 \times 10^{-3}) = 1.0 \times 10^{-3}) M$ |
| $NO_2^-$ | $0$ | $1.0 \times 10^{-3} M$ |
| $HNO_2$ | $(x) M$ | $(x - (1.0 \times 10^{-3})) M$ |

We substitute concentrations into the expression of the ionization constant:

$$K = 4.5 \times 10^{-4} = \frac{[H^+][NO_2^-]}{[HNO_2]}$$

$$4.5 \times 10^{-4} = \frac{(1.0 \times 10^{-3})^2}{x - (1.0 \times 10^{-3})}$$

$$x = 3.2 \times 10^{-3}$$

Thus, $3.2 \times 10^{-3}$ mol of $HNO_2$ must be added.

8. $[HX] = 5.0$
   [16.1]

The ionization of HX is

$$HX \rightleftharpoons H^+ + X^-$$

We make the following assumptions:

| Species | Original Concentration | Equilibrium Concentration |
|---------|------------------------|---------------------------|
| $H^+$ | $10^{-7} M$ | $(10^{-7} + 0.01x \cong 0.01x) M$ |
| $X^-$ | $0$ | |
| $HX$ | $(x) M$ | $(x - 0.01x \cong x) M$ |

We substitute concentrations into the expression for the ionization constant:

$$= 5.0 \times 10^{-4} = \frac{[H^+][X^-]}{[HX]}$$

$$= \frac{(0.01x)^2}{x}$$

$$(5.0 \times 10^{-4})x = (1.0 \times 10^{-4})x^2$$

$$x = \frac{5.0 \times 10^{-4}}{1.0 \times 10^{-4}}$$

$$= 5.0$$

Thus,

$$[HX] = 5.0$$

9. $pK = 6.1$
   [16.8]

The ionization of carbonic acid is

$$CO_2 + H_2O \rightleftharpoons H^+ + HCO_3^-$$

The expression for the ionization constant is

$$K = \frac{[H^+][HCO_3^-]}{[CO_2]}$$

a pH of 7.3 corresponds to a $[H^+]$ of $5.0 \times 10^{-8}M$:

$$-\log [H^+] = pH$$
$$= 7.3$$
$$[H^+] = 10^{-7.3}$$
$$= (10^{0.7})(10^{-8.0})$$
$$= 5.0 \times 10^{-8}M$$

Substituting concentrations into the expression for the ionization constant, we find

$$K = \frac{[H^+][HCO_3^-]}{[CO_2]}$$

$$= (5.0 \times 10^{-8})(16)$$

$$= 8.0 \times 10^{-7}$$

and

$$pK = -\log K$$
$$= -\log (8.0 \times 10^{-7})$$
$$= -0.90 + 7$$
$$= 6.1$$

We could also determine $pK$ as follows:

$$pH = pK - \log \frac{[CO_2]}{[HCO_3^-]}$$

$$pK = pH + \log \frac{[CO_2]}{[HCO_3]}$$

$$= 7.3 + \log \left(\frac{1}{16}\right)$$

$$= 7.3 + \log (6.25 \times 10^{-2})$$

$$= 7.3 + 0.796 - 2$$

$$= 6.1$$

10. pH = 7.70    Substituting concentrations into the expression for the
    [16.7, 16.8]    ionization constant, we find

$$K = \frac{[H^+][HCO_3^-]}{[CO_2]}$$

$$8.0 \times 10^{-7} = [H^+]40$$

$$[H^+] = \left(\frac{1}{40}\right)(8.0 \times 10^{-7})$$

$$= 2.0 \times 10^{-8}M$$

and

$$pH = -\log [H^+]$$

$$= -\log (2.0 \times 10^{-8})$$

$$= -0.30 + 8$$

$$= 7.70$$

11.  [16.4]

(a) pH = 9.29    We substitute concentrations into the equation relating
    pH and $\mathcal{E}$ at 25°C:

$$pH = \frac{\mathcal{E} - 0.2825}{0.05916}$$

$$= \frac{0.833 - 0.283}{0.0592}$$

$$= 9.29$$

(b) $[H^+] = 5.1 \times 10^{-10}M$  We calculate $[H^+]$:

$$pH = -\log [H^+]$$

$$[H^+] = 10^{-pH}$$

$$= 10^{-9.29}$$

$$= (10^{0.31})(10^{-10})$$

$$= 5.1 \times 10^{-10}M$$

12. pH range    The ionization of HIn is
    = 6.0 to 7.6
    [16.5]    $HIn \rightleftharpoons H^+ + In^-$

and the expression for the ionization constant is

$$K = 5.0 \times 10^{-8} = \frac{[H^+][In^-]}{[HIn]}$$

The yellow form is HIn and the blue form is In⁻. Thus, the ratio of yellow to blue is the ratio of HIn to In⁻, which is equal to 20 when the yellow color is visible. We calculate the pH corresponding to the ratio of HIn to In⁻ equal to 20:

$$K = 5.0 \times 10^{-8} = \frac{[H^+][In^-]}{[HIn]}$$

$$[H^+] = (5.0 \times 10^{-8})(2.0 \times 10^1)$$

$$= 1.0 \times 10^{-6}$$

and

$$pH = -\log [H^+]$$

$$= -\log (1.0 \times 10^{-6})M$$

$$= 6.0$$

The blue color is visible when the ratio of blue to yellow is 2, or when the ratio of In⁻ to HIn is 2. We calculate the pH corresponding to the ratio of HIn to In⁻ equal to 1/2:

$$K = 5.0 \times 10^{-8} = \frac{[H^+][In^-]}{[HIn]}$$

$$[H^+] = (5.0 \times 10^{-8})\left(\frac{1}{2}\right)$$

$$= 2.5 \times 10^{-8}M$$

and

$$pH = -\log (2.5 \times 10^{-8})$$

$$= -0.40 + 8$$

$$= 7.6$$

13. pH = 4.73
    [16.6]

The ionization of HCNO is

$$HCNO \rightleftharpoons H^+ + CNO^-$$

Since NaCNO is a strong electrolyte,

$$NaCNO \rightarrow Na^+ + CNO^-$$

We make the following assumptions:

| Species | Original Concentration | Equilibrium Concentration |
|---------|------------------------|---------------------------|
| $H^+$ | $10^{-7}M$ | $(10^{-7} + x \cong x)M$ |
| $CNO^-$ | $\dfrac{0.040 \text{ mol}}{0.250 \text{ liter}} = 0.16M$ | $(0.16 + x \cong 0.16)M$ |
| HCNO | $0.025M$ | $(0.025 - x \cong 0.025)M$ |

Substituting concentrations into the expression for the ionization constant of HCNO, we find

$$K = 1.2 \times 10^{-4} = \frac{[H^+][CNO^-]}{[HCNO]}$$

$$1.2 \times 10^{-4} = \frac{x(0.16)}{0.025}$$

$$x = 1.88 \times 10^{-5}$$

and

$$pH = -\log[H^+]$$
$$= -\log(1.88 \times 10^{-5})$$
$$= 4.73$$

14. $K = 1.0 \times 10^{-6}$
[16.1, 16.3]

The expression for the ionization constant of strychnine is

$$K = \frac{[SH^+][OH^-]}{[S]}$$

We calculate $[OH^-]$ as follows:

$$[OH^-] = \frac{1.0 \times 10^{-14}}{[H^+]}$$
$$= \frac{1.0 \times 10^{-14}}{1.0 \times 10^{-11}}$$
$$= 1.0 \times 10^{-3}$$

We make the following assumptions:

| Species | Equilibrium Concentration |
|---------|---------------------------|
| S | $(1.0 - (1.0 \times 10^{-3}) \cong 1.0)M$ |
| $SH^+$ | $1.0 \times 10^{-3}M$ (since the only source is S in the ionization) |
| $OH^-$ | $1.0 \times 10^{-3}M$ |

Substituting concentrations into the expression for the ionization constant, we find

$$K = \frac{(1.0 \times 10^{-3})^2}{1.0}$$

$$K = 1.0 \times 10^{-6}$$

15. pH = 5.0
    [16.7]

The ionization of HZ is

$$HZ \rightleftharpoons H^+ + Z^-$$

Since NaZ is a strong electrolyte,

$$NaZ \rightarrow Na^+ + Z^-$$

We make the following assumptions:

| Species | Original Concentration | Equilibrium Concentration |
|---------|------------------------|---------------------------|
| $H^+$ | $10^{-7}M$ | $(10^{-7} + x \cong x)M$ |
| HZ | $3.0 \times 10^{-1}M$ | $((3.0 \times 10^{-1}) - x \cong 3.0 \times 10^{-1})M$ |
| $Z^-$ | $1.5 \times 10^{-1}M$ | $((1.5 \times 10^{-1}) + x \cong 1.5 \times 10^{-1})M$ |

We substitute concentrations into the expression for the ionization constant:

$$K = 5.0 \times 10^{-6} = \frac{(1.5 \times 10^{-1})}{3.0 \times 10^{-1}}$$

$$x = \frac{(5.0 \times 10^{-6})(3.0 \times 10^{-1})}{1.5 \quad 10^{-1}}$$

$$= 1.0 \times 10^{-5} \text{ (Note the assumption that } x \ll 10^{-1} \text{ is valid.)}$$

Thus,

$$[H^+] = 1.0 \times 10^{-5}M$$

and

$$pH = -\log [H^+]$$

$$= -\log (1.0 \times 10^{-5})$$

$$= 5.0$$

We can also determine pH as follows:

$$pH = pK - \log \frac{[HZ]}{[Z^-]}$$

$$= -\log (5.0 \times 10^{-6}) - \log \left(\frac{0.30}{0.15}\right)$$

$$= -0.70 + 6 - 0.30$$

$$= 5.0$$

16. 0.022 mol $NH_4Cl$
    [16.6]

The ionization of $NH_3$ is

$$NH_3 + H_2O \rightleftharpoons NH_4^+ + OH^-$$

The expression for the ionization constant is

$$K = 1.8 \times 10^{-5} = \frac{[NH_4^+][OH^-]}{[NH_3]}$$

We determine $[OH^-]$ as follows:

$$pH = 10.0$$

$$[H^+] = 10^{-pH}$$

$$= 1.0 \times 10^{-10}M$$

$$[OH^-] = \frac{1.0 \times 10^{-14}}{[H^+]}$$

$$= \frac{1.0 \times 10^{-14}}{1.0 \times 10^{-10}}$$

$$= 1.0 \times 10^{-4}M$$

We make the following assumptions:

| Species | Equilibrium Concentration |
|---------|---------------------------|
| $NH_3$ | $(0.250 - (1.0 \times 10^{-4}) \cong 0.250)M$ |
| $NH_4^+$ | $(x)M$ |
| $OH^-$ | $1.0 \times 10^{-4}M$ |

We substitute concentrations into the expression for the ionization constant:

$$K = 1.8 \times 10^{-5} = \frac{[NH_4^+][OH^-]}{[NH_3]}$$

$$1.8 \times 10^{-5} = \frac{x(1.0 \times 10^{-4})}{2.50 \times 10^{-1}}$$

$$x = \frac{(1.8 \times 10^{-5})(2.50 \times 10^{-1})}{1.0 \times 10^{-14}}$$

$$= 4.5 \times 10^{-2}$$

Thus,

$$[NH_4^+] = 0.045M$$

and

$$? \text{ mol } NH_4^+ = (0.045M \ NH_4^+)(0.500 \text{ liter})$$

$$= 0.022 \text{ mol } NH_4^+$$

17. pH = 4.74
   [16.7]

The expression for the ionization constant of $HC_2H_3O_2$ is

$$K = 1.8 \times 10^{-5}) = \frac{[H^+][C_2H_3O_2^-]}{[HC_2H_3O_2]}$$

We make the following assumptions:

| Species | Original Concentration | Equilibrium Concentration |
|---|---|---|
| $H^+$ | $10^{-7}M$ | $(10^{-7} + x = x)M$ |
| $C_2H_3O_2^-$ | $1.0M$ | $\left(\dfrac{(50 \text{ ml})(1.0)}{(100 \text{ ml})} + x = 0.50 + x\right)M \cong 0.50$ |
| $HC_2H_3O_2$ | $1.0M$ | $\left(\dfrac{(50 \text{ ml})(1.0)}{(100 \text{ ml})} - x = 0.50 - x\right)M \cong 0.50$ |

We substitute concentrations into the expression for the ionization constant:

$$K = 1.8 \times 10^{-5} = \frac{[H^+][C_2H_3O_2^-]}{[HC_2H_3O_2]}$$

$$1.8 \times 10^{-5} = \frac{x(0.50)}{0.50}$$

$$x = 1.8 \times 10^{-5}$$

Thus,

$$[H^+] = 1.8 \times 10^{-5}M$$

and

$$pH = -\log[H^+]$$
$$= -\log(1.8 \times 10^{-5})$$
$$= -0.26 + 5$$
$$= 4.74$$

18. $[H^+] = 6.3 \times 10^{-5}M$   The expression for the primary ionization constant
    [16.8]                        of $H_2S$ is

$$K_1 = 1.1 \times 10^{-7} = \frac{[H^+][HS^-]}{[H_2S]}$$

We make the following assumptions:

| Species | Equilibrium Concentration |
|---|---|
| $H^+$ | $(x)M$ |
| $HS^-$ | $(x)M$ |
| $H_2S$ | $((3.6 \times 10^{-2}) - x \cong 3.6 \times 10^{-2})M$ |

We substitute concentrations into the expression for the ionization constant:

$$K = 1.1 \times 10^{-7} = \frac{[H^+][HS^-]}{[H_2S]}$$

$$1.1 \times 10^{-7} = \frac{x^2}{3.6 \times 10^{-2}}$$

$$x^2 = 40 \times 10^{-10}$$

$$x = 6.3 \times 10^{-5}$$

Thus,

$$[H^+] = 6.3 \times 10^{-5} M$$

19. $[H^+] = 2.8 \times 10^{-3} M$    The ionizations of ascorbic acid are

$[C_6H_7O_6^-] = 2.8 \times 10^{-3} M$    $C_6H_8O_6 \rightleftharpoons H^+ + C_6H_7O_6^-$

$[C_6H_6O_6^{2-}] = 1.6 \times 10^{-12} M$   $C_6H_7O_6^- \rightleftharpoons H^+ + C_6H_6O_6^{2-}$

[16.8]     For the first ionization we make the following assumptions:

| Species | Equilibrium Concentration |
|---|---|
| $C_6H_8O_6$ | $((1.0 \times 10^{-1}) - x = 1.0 \times 10^{-1}) M$ |
| $C_6H_7O_6^-$ | $(x) M$ |
| $H^+$ | $(x) M$ |

We substitute concentrations into the expression for the ionization constant of ascorbic acid $K_1$:

$$K_1 = 7.9 \times 10^{-5} = \frac{[H^+][C_6H_7O_6^-]}{[C_6H_8O_6]}$$

$$7.9 \times 10^{-5} = \frac{x^2}{1.0 \times 10^{-1}}$$

$$x^2 = 7.9 \times 10^{-6}$$

$$x = 2.8 \times 10^{-3}$$

Thus,

$$[H^+] = [C_6H_7O_6^-] = 2.8 \times 10^{-3} M$$

For the second ionization of ascorbic acid we make the following assumptions:

| Species | Equilibrium Concentration |
|---|---|
| $H^+$ | $2.8 \times 10^{-3} M$ |
| $C_6H_7O_6^-$ | $2.8 \times 10^{-3} M$ |
| $C_6H_6O_6^{2-}$ | $(y) M$ |

We substitute concentrations into the expression for the ionization constant of ascorbic acid $K_2$:

$$K_2 = 1.6 \times 10^{-12} = \frac{[H^+] [C_6H_6O_6^{2-}]}{[C_6H_7O_6^-]}$$

$$1.6 \times 10^{-12} = \frac{(2.8 \times 10^{-3}) (y)}{2.8 \times 10^{-3}}$$

$$y = 1.6 \times 10^{-12}$$

Thus,

$$[C_6H_6O_6^{2-}] = 1.6 \times 10^{-12} M$$

20. [16.7]
    (a) pH = 4.74     The ionization of $HC_2H_3O_2$ is

$$HC_2H_3O_2 \rightleftarrows H^+ + C_2H_3O_2^-$$

The expression for the ionization constant is

$$K = 1.8 \times 10^{-5} = \frac{[H^+] [C_2H_3O_2^-]}{[HC_2H_3O_2]}$$

We make the following assumptions:

| Species | Equilibrium Concentration |
|---------|---------------------------|
| $H^+$ | $(x) M$ |
| $C_2H_3O_2^-$ | $\left(\dfrac{(500 \text{ ml}) (1.0)}{1000 \text{ ml}} + x \cong 0.50\right) M$ |
| $HC_2H_3O_2$ | $\left(\dfrac{(500 \text{ ml}) (1.0)}{1000 \text{ ml}} - x \cong 0.50\right) M$ |

We substitute concentrations into the expression for the ionization constant:

$$K = 1.8 \times 10^{-5} = \frac{[H^+] [C_2H_3O_2^-]}{[HC_2H_3O_2]}$$

$$1.8 \times 10^{-5} = \frac{x(0.50)}{0.50}$$

$$x = 1.8 \times 10^{-5} \quad \text{(Note the assumption that } x \ll 0.50 \text{ is valid.)}$$

Thus,

$$[H^+] = 1.8 \times 10^{-5} M$$

and

$$\begin{aligned} pH &= -\log [H^+] \\ &= -\log (1.8 \times 10^{-5}) \\ &= -0.26 + 5 \\ &= 4.74 \end{aligned}$$

(b) pH = 4.73    The $H^+$ from the completely dissociated HCl will react with $C_2H_3O_2^-$ to form $HC_2H_3O_2$:

$$H^+ + C_2H_3O_2 \rightleftharpoons HC_2H_3O_2$$

The protonation of $C_2H_3O_2^-$ has a $K$ equal to $1/(1.8 \times 10^{-5})$, or $5.6 \times 10^4$. Since this constant is large, we assume that all the $H^+$ reacts to form $HC_2H_3O_2$. Thus, we make the following assumptions:

| Species | Equilibrium Concentration |
|---------|---------------------------|
| $H^+$ | $(y)M$ |
| $C_2H_3O_2^-$ | $(0.50 - 0.005 = 0.495)M$ |
| $HC_2H_3O_2$ | $(0.50 + 0.005 = 0.505)M$ |

We substitute concentrations into the equation relating pH and p$K$:

$$pH = pK - \log \frac{[HC_2H_3O_2]}{[C_2H_3O_2^-]}$$

$$= 4.74 - \log \left(\frac{0.505}{0.495}\right)$$

$$= 4.74 - \log 1.02$$

$$= 4.74 - 0.009$$

$$= 4.73$$

The buffer solution closely maintains its pH in spite of the addition of the hydrochloric acid.

SELF-TEST

Complete the test in 30 minutes:

1. What is the pH of a solution that has an $OH^-$ concentration of $0.075M$?

2. What is the $H^+$ concentration of a solution with a pH of 9.6?

3. A weak acid, HX, is 0.20 percent ionized in $0.50M$ solution. What is the ionization constant of the acid?

4. The ionization constant of a weak acid HX is $6.0 \times 10^{-6}$. What concentration of NaX must be present in a $0.30M$ solution of HX to prepare a buffer with a pH of 5.00?

# Ionic Equilibria, Part II

OBJECTIVES

I. You should be able to demonstrate your knowledge of the following terms by defining them, describing them, or giving specific examples of them:

amphoteric [17.5]
complex ion [17.4]
equivalence point [17.7]
formation constant [17.4]
hydrolysis [17.6]
instability constant [17.4]
ion product [17.2]
$K_a$ [17.6]
$K_b$ [17.6]
$K_{SP}$ [17.1]
ligand [17.4]
salt effect [17.1]
saturated [17.2]
stability constant [17.4]

II. You should be able to determine $K_{SP}$ from experimental data and to calculate the solubility of a compound from the value of $K_{SP}$.

III. You should be able to determine the concentrations of species in a solution that contains complex ions.

IV. You should be able to determine the pH of a solution containing a known concentration of a compound that will hydrolyze.

V. You should be able to predict the shape of titration curves.

EXERCISES

I. Answer each of the following questions without performing any calculations. This exercise will increase your qualitative understanding of equilibria. You may refer to equilibrium constant data if necessary.

1. The pH of pure water at 25°C is closest to
(a) 6.0          (d) 7.5
(b) 6.5          (e) 8.0
(c) 7.0

2. The pH of a $10^{-6}M$ HCl is closest to
(a) 6.0          (d) 7.5
(b) 6.5          (e) 8.0
(c) 7.0

3. The pH of a $10^{-8}M$ HCl solution is closest to
(a) 6.0          (d) 7.5
(b) 6.5          (e) 8.0
(c) 7.0

4. The pH of a $10^{-3}M$ NaOH solution is closest to
(a) 3.0          (c) 9.0
(b) 5.0          (d) 11.0

5. The pH of a $10^{-3}M$ NaCl solution is closest to
(a) 3.0          (d) 8.0
(b) 6.0          (e) 11.0
(c) 7.0

6. The pH of a $10^{-3}M$ acetic acid ($K = 1.8 \times 10^{-5}$) is closest to
   (a) 1.0
   (b) 4.0
   (c) 7.0
   (d) 10.0
   (e) 13.0

7. The species in a $10^{-3}M$ acetic acid solution that has the highest concentration is
   (a) $H^+$
   (b) $OH^-$
   (c) $Na^+$
   (d) $C_2H_3O_2^-$
   (e) $HC_2H_3O_2$

8. The species in a $10^{-5}M$ HCl solution that has the highest concentration is
   (a) $H^+$
   (b) $OH^-$
   (c) $Na^+$
   (d) $Cl^-$
   (e) HCl

9. Five solutions are prepared. Each solution is $10^{-2}M$ in one of the following. Which solution is the most acidic?
   (a) $HClO_2$, $K = 1.1 \times 10^{-2}$
   (b) $HC_2H_3O_2$, $K = 1.8 \times 10^{-5}$
   (c) HCN, $K = 4.0 \times 10^{-10}$
   (d) HF, $K = 6.7 \times 10^{-4}$
   (e) $NH_3$, $K = 1.8 \times 10^{-5}$

10. A solution that is $10^{-2}M$ in acetic acid and $10^{-2}M$ in sodium acetate is prepared. The p$K$ of acetic acid is 4.7. The pH of the solution is closest to
    (a) 2.0
    (b) 4.7
    (c) 5.0
    (d) 5.7
    (e) 7.0

11. The most soluble of the following carbonates is
    (a) $BaCO_3$, $K_{SP} = 1.6 \times 10^{-9}$
    (b) $CdCO_3$, $K_{SP} = 5.2 \times 10^{-12}$
    (c) $CaCO_3$, $K_{SP} = 4.7 \times 10^{-9}$
    (d) $PbCO_3$, $K_{SP} = 1.5 \times 10^{-15}$
    (e) $MgCO_3$, $K_{SP} = 1 \times 10^{-15}$

12. The $K_{SP}$ of $CdCO_3$ is $5.2 \times 10^{-12}$, and the solubility of $CdCO_3$ calculated with that value is $2.6 \times 10^{-6}$ mol/liter. The experimentally determined solubility is
    (a) much lower because of temperature effects
    (b) much higher because of carbonate hydrolysis
    (c) much higher because of temperature effects
    (d) much higher because of supersaturation
    (e) the same

13. Which of the following indicators would be best for the titration of 0.1$M$ HNO$_2$ with 0.1$M$ NaOH?
    (a) methyl orange, which has a pH range for color change of 3.1 to 4.5
    (b) methyl red, which has a pH range for color change of 4.2 to 6.3
    (c) litmus, which has a pH range for color change of 5.0 to 8.0
    (d) thymol blue, which has a pH range for color change of 8.0 to 9.6
    (e) all of the preceding

14. The value of the solubility constant of silver(I) sulfate is approximately the same as that of calcium sulfate. Which of the following statements is true?
    (a) Calcium sulfate is more soluble than silver sulfate.
    (b) Silver sulfate is more soluble than calcium sulfate.
    (c) Silver sulfate is as soluble as calcium sulfate.
    (d) Nothing can be said about the relative solutilities of silver sulfate and calcium sulfate.
    (e) None of the preceding is true.

15. The pH of a solution is 4.50. What is the pOH of the solution?
    (a) 4.50          (d) 8.50
    (b) 5.50          (e) 9.50
    (c) 6.50

16. How many milliliters of 0.05$M$ NaOH are needed to titrate 50 ml of 0.1$M$ acetic acid?
    (a) 25          (d) 200
    (b) 50          (e) 500
    (c) 100

17. The pH of 10$^{-3}M$ NH$_4$C$_2$H$_3$O$_2$ is closest to
    (a) 3          (d) 9
    (b) 5          (e) 11
    (c) 7

II. Answer the following:

1. What is the pH of a 0.10$M$ NaNO$_2$ solution? The ionization constant of HNO$_2$, a weak acid, is $4.5 \times 10^{-4}$.

2. At 25°C a saturated solution of Ca(OH)$_2$ has a pH of 12.45.
    (a) What is the molar solubility of Ca(OH)$_2$?
    (b) What is the $K_{SP}$ of Ca(OH)$_2$?

3. A 0.10$M$ solution of KCN has a pH of 11.00. What is the ionization constant, $K$, of HCN?

4. What is the pH of a 0.10$M$ Na$_2$S solution? The ionization constant for the primary ionization of H$_2$S is $1.1 \times 10^{-7}$ and that for the secondary ionization is $1.0 \times 10^{-14}$.

5. Calculate the $K_{SP}$ of lead iodide from the following standard electrode potentials:

$$2e^- + PbI_2(s) \rightleftarrows Pb(s) + 2I^-(aq) \qquad \mathscr{E}° = -0.365 \text{ V}$$
$$2e^- + Pb^{2+}(aq) \rightleftarrows Pb(s) \qquad \mathscr{E}° = -0.126 \text{ V}$$

6. Will a precipitate form in a solution that is $1.00 \times 10^{-3}M$ in Ba$^{2+}$ and $1.00 \times 10^{-3}M$ in SO$_4^{2-}$. The $K_{SP}$ of BaSO$_5$ is $1.5 \times 10^{-9}$.

7. How many moles of Ag$_2$C$_2$O$_4$ will dissolve in 100 ml of 0.050$M$ Na$_2$C$_2$O$_4$? Neglect the hydrolysis of ions. The $K_{SP}$ of Ag$_2$C$_2$O$_4$ is $1.1 \times 10^{-11}$.

8. Which is the least soluble: AgCl or Ag$_2$CrO$_4$? The $K_{SP}$ of AgCl is $1.7 \times 10^{-10}$ and that of Ag$_2$CrO$_4$ is $1.0 \times 10^{-12}$.

9. A solution is 0.20$M$ in Ca$^{2+}$ and 0.10$M$ in Sr$^{2+}$.
   (a) If solid Na$_2$SO$_4$ is added very slowly to the solution, which will precipitate first: CaSO$_4$ or SrSO$_4$? Neglect volume changes. The $K_{SP}$ of CaSO$_4$ is $2.4 \times 10^{-5}$ and that of SrSO$_4$ is $7.6 \times 10^{-7}$.
   (b) The addition of Na$_2$SO$_4$ is continued until the second cation starts to precipitate as the sulfate. What is the concentration of the first cation at that point?

10. What are the final concentrations of Fe$^{2+}$ and Cd$^{2+}$ in a solution originally 0.10$M$ in Fe$^{2+}$ and 0.10$M$ in Cd$^{2+}$, which is buffered with 1.0$M$ SO$_4^{2-}$ and 1.0$M$ HSO$_4^-$ and saturated with H$_2$S? For a saturated solution of H$_2$S

$$[H^+]^2[S^{2-}] = 1.1 \times 10^{-22}$$

The ionization constant for the secondary ionization of H$_2$SO$_4$ is $1.3 \times 10^{-2}$. The $K_{SP}$ of FeS is $4.0 \times 10^{-19}$ and that of CdS is $1.0 \times 10^{-28}$.

11. A solution is prepared by mixing 50 ml of 0.020$M$ Ag$^+$ with 50 ml of 0.20$M$ NH$_3$. The principal equilibrium is

$$Ag^+ + 2NH_3 \rightleftarrows Ag(NH_3)_2^+$$

and the equilibrium expression is

$$K = \frac{[Ag(NH_3)_2^+]}{[Ag^+][NH_3]^2} = 1.67 \times 10^7$$

What are the final concentrations of $Ag^+$, $NH_3$, $Ag(NH_3)_2^+$, $H^+$, $OH^-$, $NH_3$, and $NH_4^+$ in the solution? The equilibrium constant for the hydrolysis of $NH_3$ is $1.8 \times 10^{-5}$.

12. Glutaramic acid has an ionization constant of $4.0 \times 10^{-5}$. Draw the titration curve for the titration of 50 ml of $0.200N$ glutaramic acid with $0.200N$ NaOH. The $0.200N$ glutaramic acid solution is diluted to 100 ml before any base is added.

## ANSWERS TO EXERCISES

I. Concepts of ionic equilibria

1. 7.0
   [16.3]

Since

$$K = 1.0 \times 10^{-14} = [H^+][OH^-]$$

and in pure water

$$[H^+] = [OH^-]$$

we determine the pH to be equal to 7:

$$[H^+][OH^-] = 1.0 \times 10^{-14}$$
$$[H^+]^2 = 1.0 \times 10^{-14}$$
$$[H^+] = 1.0 \times 10^{-7}$$
$$pH = -\log [H^+] = 7$$

2. 6.0
   [16.3]

Since HCl completely dissociates,

$$[H^+] \cong 10^{-6}M$$

Therefore,

$$pH = 6.0$$

3. 7.0
   [16.3]

The $[H^+]$ from HCl is $10^{-8}M$, but that from the dissociation of water is much greater, $\sim 10^{-7}M$. Thus, pH = 7.0. The pH is slightly less than 7 because of the presence of $[H^+]$ from HCl.

4. 11.0
   [16.3]

Since NaOH completely dissociates

$$NaOH \rightarrow Na^+ + OH^-$$

We determine the pH as follows:

$$pOH = -\log [OH^-] = 3$$
$$pH = 14 - 3$$
$$pH = 11.0$$

5. 7.0
   [16.3]

A $10^{-3}M$ NaCl solution is neutral.

6. 4.0
   [16.7]

If acetic acid were a strong acid, i.e., if it were completely dissociated in water, the pH would be 3. If it were not dissociated at all, the pH would be 7. Acetic acid actually lies somewhere between the limits; thus, 4.0 is the best choice.

7. $HC_2H_3O_2$
   [16.1]

In solutions of weak acids the most concentrated species is usually the acid.

8. $H^+$
   [16.3]

In water HCl completely dissociates, giving $10^{-5}M$ $H^+$ and $10^{-5}M$ $Cl^-$. There is a small contribution of $H^+$ from the ionization of water; thus, $[H^+]$ is slightly greater than $[Cl^-]$.

9. $HClO_2$
   [16.1]

A larger dissociation constant corresponds to a greater degree of dissociation.

10. 4.7
    [16.6]

The pH equals the p$K$ since $[HC_2H_3O_2]$ equals $[C_2H_3O_2^-]$.

$$pH = pK + \log \frac{[C_2H_3O_2^-]}{[HC_2H_3O_2]}$$

$$= 4.7 + 0$$

$$= 4.7$$

11. $CaCO_3$
    [17.1]

The most soluble of the carbonates is the one with the largest value of $K_{SP}$:

$$K_{SP} = [M^{2+}][CO_3^{2-}]$$

The largest value of $K_{SP}$ corresponds to the compound that gives the most ions in solution.

12. much higher because of carbonate hydrolysis
    [17.6]

13. thymol blue
    [17.7]

At the equivalence point the number of moles of acid is equal to the number of moles of base. Since $HNO_2$ is a weak acid, the conjugate base $NO_2^-$ will hydrolyze, thus producing a basic solution. The indicator must therefore change color in the basic region.

14. Calcium sulfate is more soluble than silver sulfate.
    [17.1]

If we indicate molar solubility by $S$, we find that

$$K_{SP} = S(2S)^2$$

$$= 4S^3$$

and for $CaSO_4$

$$K_{SP} = S^2$$

If the value of $K_{SP}$ is the same for both compounds, we find that for $Ag_2SO_4$

$$S = \sqrt[3]{(1/4)K_{SP}}$$

and for $CaSO_4$

$$S = \sqrt{K_{SP}}$$

Thus, the solubility of $CaSO_4$ is larger.

15. 9.50
   [16.3]

We determine pOH as follows:

$$pH + pOH = 14.00$$

$$pOH = 14.00 - pH$$

$$= 14.00 - 4.50$$

$$= 9.50$$

16. 100 ml
   [17.7]

The number of moles of acid equals the number of moles of bases:

$$moles \ of \ acid = moles \ of \ base$$

$$(VM)_{acid} = (VM)_{base}$$

$$(0.050 \ liter)\left(\frac{0.1 \ mol}{1 \ liter}\right) = V_{base}\left(\frac{0.05 \ mol}{1 \ liter}\right)$$

$$V_{base} = 100 \ ml$$

17. 7°
   [17.6]

II. Ionic equilibria calculations

1. pH = 8.17
   [17.6]

Sodium nitrite is a strong electrolyte. Thus,

$$NaNO_2(s) \rightarrow Na^+(aq) + NO_2^-(aq)$$

The hydrolysis of $NO_2^-$ is

$$NO_2^-(aq) + H_2O \rightleftharpoons HNO_2(aq) + OH^-(aq)$$

We obtain the equilibrium constant for the hydrolysis as follows:

$$K_H = \frac{[HNO_2][OH^-]}{[NO_2^-]} = \frac{K_W}{K_{HNO_2}} = \frac{1.0 \times 10^{-14}}{4.5 \times 10^{-4}}$$

$$= 2.22 \times 10^{-11}$$

We make the following assumptions:

| Species | Equilibrium Concentration |
|---------|---------------------------|
| $HNO_2$ | $(x)M$ |
| $OH^-$ | $(x)M$ |
| $NO_2^-$ | $(0.10 - x \cong 0.10)M$ |

We substitute concentrations into the equilibrium expression for the hydrolysis of $NO_2^-$:

$$K_H = 2.22 \times 10^{-11} = \frac{[HNO_2][OH^-]}{[NO_2^-]}$$

$$= \frac{x^2}{0.10}$$

$$x^2 = 2.22 \times 10^{-22}$$

$$x = 1.49 \times 10^{-6}$$

Thus,

$$[OH^-] = 1.49 \times 10^{-6}$$

We calculate pH as follows:

$$[H^+][OH^-] = 1.0 \times 10^{-14}$$

$$[H^+] = \frac{1.0 \times 10^{-14}}{[OH^-]}$$

$$= \frac{1.0 \times 10^{-14}}{1.49 \times 10^{-6}}$$

$$= 6.71 \times 10^{-9}$$

$$pH = -\log (6.71 \times 10^{-9})$$

$$= 8.17$$

2. [17.1]

(a) $K_{SP} = 1.1 \times 10^{-5}$ The equilibrium between solid $Ca(OH)_2$ and a saturated solution of $Ca(OH)_2$ is

$$Ca(OH)_2(s) \rightleftharpoons Ca^{2+}(aq) + 2OH^-(aq)$$

and the $K_{SP}$ is

$$K_{SP} = [Ca^{2+}][OH^-]^2$$

We calculate the $[OH^-]$ as follows:

$$pH + pOH = 14$$

$$pOH = 14 - pH$$

$$= 14 - 12.45$$

$$= 1.55$$

$$[OH^-] = 2.82 \times 10^{-2} M$$

According to the balanced equation 2 moles of $OH^-$ are produced for every mole of $Ca^{2+}$. Thus,

$$[Ca^{2+}] = \frac{[OH^-]}{2}$$

$$= \frac{(2.82 \times 10^{-2})}{2}$$

$$= 1.41 \times 10^{-2}$$

We calculate the $K_{SP}$:

$$K_{SP} = [Ca^{2+}][OH^-]^2$$

$$= (1.41 \times 10^{-2})(2.82 \times 10^{-2})^2$$

$$= 1.1 \times 10^{-5}$$

(b) $1.4 \times 10^{-2}$    The molar solubility of $Ca(OH)_2$ at 25°C is the number of moles of $Ca(OH)_2$ dissolved per liter of solution at that temperature. Since 1 mole of $Ca^{2+}$ is present per mole of $Ca(OH)_2$ dissolved,

$$\text{molar solubility of } Ca(OH)_2 = [Ca^{2+}]$$

$$= 1.41 \times 10^{-2}$$

3. $K = 1.0 \times 10^{-9}$
[17.6]

Potassium cyanide is a strong electrolyte:

$$KCN \rightarrow K^+ + CN^-$$

The cyanide ion is hydrolyzed:

$$CN^- + H_2O \rightleftharpoons HCN + OH^-$$

The equilibrium expression for the hydrolysis is

$$K_H = \frac{K_W}{K_{HCN}} = \frac{[HCN][OH^-]}{[CN^-]}$$

We calculate the $[OH^-]$:

$$pOH = 14 - pH$$

$$= 14 - 11.00$$

$$= 3.00$$

$$[OH^-] = 10^{-pOH}$$

$$= 1.00 \times 10^{-3}$$

We make the following assumptions:

| Species | Equilibrium Concentration |
|---------|---------------------------|
| $OH^-$ | $1.00 \times 10^{-3} M$ |
| HCN | $1.00 \times 10^{-3} M$ |
| $CN^-$ | $(1.00 \times 10^{-1} M) - (1.0 \times 10^{-3} M) \cong 1.0 \times 10^{-1} M$ |

We substitute concentrations into the expression for the $K_H$:

$$K_H = \frac{K_W}{K_{HCN}} = \frac{[HCN][OH^-]}{[CN^-]}$$

$$\frac{1.0 \times 10^{-14}}{K_{HCN}} = \frac{(1.00 \times 10^{-3})(1.00 \times 10^{-3})}{1.0 \times 10^{-1}}$$

$$K_{HCN} = \frac{(1.0 \times 10^{-14})(1.0 \times 10^{-1})}{(1.00 \times 10^{-3})^2}$$

$$= 1.0 \times 10^{-9}$$

4. pH = 12.96
   [17.6]

We consider the following equilibria:

$$S^{2-} + H_2O \rightleftharpoons HS^- + OH^- \qquad K_H = \frac{K_W}{K_{2H_2S}} = \frac{1.0 \times 10^{-14}}{1.0 \times 10^{-14}} = 1.00$$

$$HS^- + H_2O \rightleftharpoons H_2S + OH^- \qquad K_H = \frac{K_W}{K_{1H_2S}} = \frac{1.0 \times 10^{-14}}{1.1 \times 10^{-7}} = 9.09 \times 10^{-8}$$

$$H_2O \rightleftharpoons H^+ + OH^- \qquad K_W = 1.0 \times 10^{-14}$$

Since the $K$ of the first equilibrium is much larger than either of the other two, the first equilibrium is the most important and the only one we must consider.

The expression for the hydrolysis of $S^{2-}$ is

$$K_H = 1.00 = \frac{[HS^-][OH^-]}{[S^{2-}]}$$

We make the following assumptions:

| Species | Equilibrium Concentration |
|---------|---------------------------|
| $OH^-$ | $(x)M$ |
| $HS^-$ | $(x)M$ |
| $S^{2-}$ | $(0.10 - x)M$ |

We substitute concentrations into the expression for the hydrolysis constant of $S^{2-}$:

$$K_H = 1.00 = \frac{[HS^-][OH^-]}{[S]^{2-}}$$

$$1.00 = \frac{x^2}{(0.10 - x)}$$

$$x^2 = 0.100 - 1.00$$

$$x^2 + 1.00x - 0.100 = 0$$

We use the quadratic formula to solve for

$$x = \frac{-1.00 \pm \sqrt{(1.00)^2 - 4(1.00)(-1.100)}}{2(1.00)}$$

$$x = 9.16 \times 10^{-2} = OH^-$$

The pH is then calculated:

$$[H^+] = \frac{1.0 \times 10^{-14}}{9.16 \times 10^{-2}}$$

$$= 1.09 \times 10^{-13}$$

$$pH = 12.96$$

5. $K_{SP} = 8.1 \times 10^{-9}$  The equilibrium between solid lead iodide and a saturated
   [17.1]                      solution of lead iodide is

$$PbI_2(s) \rightleftharpoons Pb^{2+}(aq) + 2I^-(aq)$$

We calculate the $\mathcal{E}^\circ$ of the cell from the half reactions:

$$
\begin{array}{lll}
2e^- + PbI_2(s) \rightleftharpoons Pb(s) + 2I^-(aq) & \mathcal{E}^\circ = -0.365 \text{ V} \\
\underline{Pb(s) \rightleftharpoons Pb^{2+}(aq) + 2e^-} & \underline{\mathcal{E}^\circ = +0.126 \text{ V}} \\
PbI_2(s) \rightleftharpoons Pb^{2+}(aq) + 2I^-(aq) & \mathcal{E}^\circ = -0.239 \text{ V}
\end{array}
$$

We calculate $K_{SP}$ as follows:

$$\Delta G^\circ = -nF\mathcal{E}^\circ$$

$$= -(2)(9.65 \times 10^4 \text{ J/V})(-0.239 \text{ V})$$

$$= 4.61 \times 10^4 \text{ J}$$

$$\Delta G^\circ = -2.303RT \log K$$

$$\log K = -\frac{\Delta G^\circ}{2.303RT}$$

$$= -\frac{(4.61 \times 10^4 \text{ J})}{(2.303)(8.31 \text{ J/K·mol})(298 \text{ K})}$$

$$= -8.09$$

$$K = 8.1 \times 10^{-9}$$

6. yes       The $K_{SP}$ of $BaSO_4$ is
   [17.2]

$$K_{SP} = 1.5 \times 10^{-9} = [Ba^{2+}][SO_4^{2-}]$$

In a solution that is $0.001M$ in $Ba^{2+}$ and $0.001M$ in $SO_4^{2-}$,
the ion product is

$$\text{ion product} = [Ba^{2+}][SO_4^{2-}]$$

$$= (1 \times 10^{-3})(1 \times 10^{-3})$$

$$= 1 \times 10^{-6}$$

If the ion product is greater than $K_{SP}$, precipitation
will occur until the ion product equals the $K_{SP}$.
Since the ion product of $BaSO_4$ is greater than the $K_{SP}$,
precipitation will occur.

7. $7.4 \times 10^{-7}$ mol   Sodium oxalate is a strong electrolyte:
   [17.2]

$$Na_2C_2O_4 \rightarrow 2Na^+ + C_2O_4^{2-}$$

In a saturated solution the ion product is equal to the
$K_{SP}$ and no more solid will dissolve. The equilibrium between solid $Ag_2C_2O_4$ and a saturated solution of $Ag_2C_2O_4$ is

$$Ag_2C_2O_4 \rightleftharpoons 2Ag^+ + C_2O_4^{2-}$$

and the $K_{SP}$ for $Ag_2C_2O_4$ is

$$K_{SP} = 1.1 \times 10^{-11} = [Ag^+]^2[C_2O_4^{2-}]$$

We make the following assumptions:

| Species | Equilibrium Concentration |
|---|---|
| $Ag^+$ | $(2x)M$ (where $x$ = molar solubility of $Ag_2C_2O_4$) |
| $C_2O_4^{2-}$ | |
| $C_2O_4^{2-}$ | $((5.0 \times 10^{-2}) + x \cong 5.0 \times 10^{-2})M$ |

We substitute concentrations into the expression for the $K_{SP}$:

$$K_{SP} = 1.1 \times 10^{-11} = [Ag]^2[C_2O_4^{2-}]$$
$$= (2x)^2(5.0 \times 10^{-2})$$
$$= (2.0 \times 10^{-1})(x^2)$$
$$x^2 = \frac{1.1 \times 10^{-11}}{2.0 \times 10^{-1}}$$
$$x^2 = 5.5 \times 10^{-11}$$
$$x = 7.4 \times 10^{-6}$$

Thus, $7.4 \times 10^{-6}$ mol of $Ag_2C_2O_4$ will dissolve per liter of solution. We calculate the number of moles that will dissolve in 100 ml of solution:

$$? \text{ mol} = \left(\frac{7.4 \times 10^{-6} \text{ mol}}{1000 \text{ ml}}\right)100 \text{ ml}$$
$$= 7.4 \times 10^{-7} \text{ mol}$$

8. AgCl
   [17.2]

The $K_{SP}$ of AgCl is $1.7 \times 10^{-10}$ and that of $Ag_2CrO_4$ is $1.9 \times 10^{-12}$. The compound with the smaller molar solubility is the less soluble. We calculate the molar solubility of each compound.

For AgCl, the equilibrium between solid silver chloride and a saturated solution of silver chloride is

$$AgCl \rightleftharpoons Ag^+ + Cl^-$$

Since 1 mol of $Ag^+$ and 1 mol of $Cl^-$ are produced per mole of AgCl dissolved, the molar solubility of AgCl, $x$, is

$$x = [Ag^+] = [Cl^-]$$

We calculate $x$ from the $K_{SP}$ of AgCl:

$$K_{SP} = 1.7 \times 10^{-10} - [Ag^+][Cl^-]$$
$$1.7 \times 10^{-10} = x^2$$
$$x = 1.3 \times 10^{-5}$$

For $Ag_2CrO_4$, the equilibrium between solid silver chromate and a saturated solution of silver chromate is

$$Ag_2CrO_4 \rightleftharpoons 2Ag^+ + CrO_4^{2-}$$

Since 2 mol of $Ag^+$ and 1 mol of $CrO_4^{2-}$ are produced per 1 mol of $Ag_2CrO_4$ dissolved, the molar solubility of $Ag_2CrO_4$, $y$, is

$$y = \frac{[Ag^+]}{2} = [CrO_4^{2-}]$$

We calculate $y$ from the $K_{SP}$ of $Ag_2CrO_4$:

$$K_{SP} = 1.9 \times 10^{-12} = [Ag^+]^2[CrO_4]$$

$$= [2y]^2[y]$$

$$= 4y^3$$

$$y = 7.8 \times 10^{-5}$$

Thus, AgCl is the less soluble.

9. [17.2]
   (a) $SrSO_4$

The $K_{SP}$ of $CaSO_4$ is $2.4 \times 10^{-5}$ and that of $SrSO_4$ is $7.6 \times 10^{-7}$:

$$K_{SP} = 2.4 \times 10^{-5} = [Ca^{2+}][SO_4^{2-}]$$

$$K_{SP} = 7.6 \times 10^{-7} = [Sr^{2+}][SO_4^{2-}]$$

Precipitation begins when the ion product just exceeds the $K_{SP}$. A greater concentration of $SO_4^{2-}$ is necessary to exceed the $K_{SP}$ of $CaSO_4$ than to exceed that of $SrSO_4$.

   (b) $[Sr^{2+}] = 6.3$
       $\times 10^{-3}M$

We know that $[SO_4^{2-}] = 1.2 \times 10^{-4}M$ when $Ca^{2+}$ begins to precipitate as $CaSO_4$:

$$[Ca^{2+}][SO_4^{2-}] = 2.4 \times 10^{-5}$$

$$[SO_4^{2-}] = \frac{2.4 \times 10^{-5}}{[Ca^{2+}]}$$

$$= \frac{2.4 \times 10^{-5}}{2.1 \times 10^{-1}}$$

$$= 1.2 \times 10^{-4}$$

We can determine $[Sr^{2+}]$ when $Ca^{2+}$ begins to precipitate as follows:

$$K_{SP} = 7.6 \times 10^{-7} = [Sr^{2+}][SO_4^{2-}]$$

$$[Sr^{2+}] = \frac{7.6 \times 10^{-7}}{1.2 \times 10^{-4}}$$

$$= 6.3 \times 10^{-3}$$

10. $[Fe^{2+}] = 0.10M$
    $[Cd^{2+}] = 1.5$
    $\times 10^{-10}M$
    [17.2]

To determine whether a metal sulfide precipitates, we calculate the ion product for each compound. First we determine $[S^{2-}]$ in a saturated $H_2S$ solution:

$$[H^+]^2[S^{2-}] = 1.1 \times 10^{-22}$$

The $[H^+]$ is controlled by the $HSO_4^-/SO_4^{2-}$ buffer:

$$HSO_4^- \rightleftharpoons H^+ + SO_4^{2-}$$

Thus, we calculate $[H^+]$ from the ionization expression for $HSO_4^-$:

$$K = 1.3 \times 10^{-2} = \frac{[H^+][SO_4^{2-}]}{[HSO_4^-]}$$

We make the following assumptions:

| Species | Equilibrium Concentration |
|---------|---------------------------|
| $H^+$ | $(x)M$ |
| $SO_4^{2-}$ | $(1.0 + x \cong 1.0)M$ |
| $HSO_4^-$ | $(1.0 - x \cong 1.0)M$ |

We substitute concentrations into the expression for the ionization constant of $HSO_4^-$:

$$K = 1.3 \times 10^{-2} = \frac{[H^+][SO_4^{2-}]}{[HSO_4^-]}$$

$$= \frac{(x)(1.0)}{1.0}$$

$$x = 1.3 \times 10^{-2}$$

Thus, $[H^+] - 1.3 \times 10^{-2}M$.

We determine the $[S^{2-}]$:

$$[H^+]^2[S^{2-}] = 1.1 \times 10^{-22}$$

$$[S^{2-}] = \frac{1.1 \times 10^{-22}}{[H^+]^2}$$

$$= \frac{1.1 \times 10^{-22}}{(1.3 \times 10^{-2})^2}$$

$$= 6.51 \times 10^{-19}$$

The ion product just exceeds $K_{SP}$ when the metal sulfide begins to precipitate. Thus, for FeS the ion product is

$$[Fe^{2+}][S^{2-}] = (1.0 \times 10^{-1})(6.5 \times 10^{-19})$$

$$= 6.51 \times 10^{-20}$$

and the $K_{SP}$ is $4.0 \times 10^{-19}$. Since the $K_{SP}$ of FeS is greater than the ion product, FeS will not precipitate and the final $[Fe^{2+}]$ is $0.10M$.

For CdS the ion product is

$$[Cd^{2+}][S^{2-}] = (1.0 \times 10^{-1})(6.5 \times 10^{-19})$$
$$= 6.5 \times 10^{-20}$$

and the $K_{SP}$ is $1.0 \times 10^{-28}$.

Since the $K_{SP}$ of CdS is less than the ion product, CdS will precipitate:

$$Cd^{2+} + H_2S \rightarrow 2H^+ + CdS$$

We calculate the concentration of $Cd^{2+}$ after precipitation of CdS:

$$K_{SP} = 1.0 \times 10^{-28} = [Cd^{2+}][S^{2-}]$$

$$[Cd^{2+}] = \frac{1.0 \times 10^{-28}}{[S^{2-}]}$$

$$= \frac{1.0 \times 10^{-28}}{6.5 \times 10^{-19}}$$

$$= 1.5 \times 10^{-10}$$

11. $[Ag^+] = 9.4 \times 10^{-8} M$

$[NH_3] = 0.08 M$

$[Ag(NH_3)_2^+] = 1.0 \times 10^{-2} M$

$[NH_4^+] = 1.2 \times 10^{-3} M$

$[OH^-] = 1.2 \times 10^{-3} M$

$[H^+] = 8.3 \times 10^{-12} M$

[17.4]

We make the following assumptions:

| Species | Equilibrium Concentration |
|---------|---------------------------|
| $Ag^+$ | $(x) M$ |
| $NH_3$ | $(0.10 - .02 + 2x \cong 0.08) M$ |
| $Ag(NH_3)_2^+$ | $(0.010 - x \cong 0.010) M$ |

We substitute concentrations into the expression for the formation constant of $Ag(NH_3)_2^+$:

$$K = 1.67 \times 10^7 = \frac{[Ag(NH_3)_2^+]}{[Ag^+][NH_3]^2}$$

$$= \frac{0.01}{(x)(0.08)^2}$$

$$x = \frac{0.010}{(0.08)^2(1.67 \times 10^7)}$$

$$= 9.36 \times 10^{-8}$$

Thus,

$$[NH_3] = 0.08 M - 2(8.36 \times 10^{-8} M)$$

$$= 0.08$$

$$[Ag(NH_3)_2^+] = (0.010 M) - (9.36 \times 10^{-8} M)$$

$$= 0.010 M$$

$$[Ag^+] = 9.36 \times 10^{-8} M$$

We use the equilibrium expression for the hydrolysis of $NH_3$ to determine $[OH^-]$ and $[NH_4^+]$:

$$NH_3(aq) + H_2O \rightleftharpoons NH_4^+(aq) + OH^-(aq)$$

$$K_H = 1.8 \times 10^{-5} = \frac{[NH_4^+][OH^-]}{[NH_3]}$$

We make the following assumptions:

| Species | Equilibrium Concentration |
|---------|---------------------------|
| $OH^-$ | $(x)M$ |
| $NH_4^+$ | $(x)M$ |
| $NH_3$ | $(0.08 - x \cong 0.08)M$ |

We substitute concentrations into the expression for the hydrolysis constant of $NH_3$:

$$K_H = 1.8 \times 10^{-5} = \frac{[NH_4^+][OH^-]}{[NH_3]}$$

$$= \frac{x^2}{0.08}$$

$$x^2 = 1.44 \times 10^{-6}$$

$$x = 1.20 \times 10^{-3}$$

Thus,

$$[NH_3] = 0.08M$$
$$[NH_4^+] = 1.2 \times 10^{-3} M$$
$$[OH^-] = 1.2 \times 10^{-3} M$$

We determine $[H^+]$:

$$[H^+][OH^-] = 1.0 \times 10^{-14}$$

$$[H^+] = \frac{1.0 \times 10^{-14}}{1.20 \times 10^{-3}}$$

$$= 8.33 \times 10^{-12}$$

Thus,

$$[H^+] = 8.3 \times 10^{-12} M$$

12. [17.7]     The titration curve is

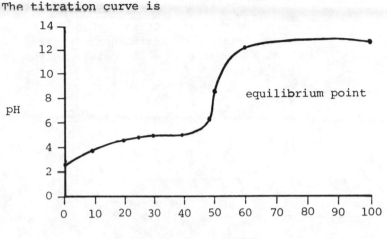

ml NaOH

We determine the titration curve for the titration of glutaramic acid, a weak acid, with sodium hydroxide, a strong base, as follows:

For simplicity we will use HA to designate glutaramic acid. At the initial point of the titration curve (i.e., the point at which no base has been added) the solution contains only HA, $H^+$, and $A^-$. We make the following assumptions:

| Species | Equilibrium Concentration |
|---------|---------------------------|
| $H^+$ | $(x)M$ |
| $A^-$ | $(x)M$ |
| HA | $\left(\dfrac{(50.0 \text{ ml})(0.200M)}{100 \text{ ml}} - x\right)M \cong 0.100M$ |

We substitute concentrations into the expression for the ionization constant of HX:

$$K = 4.0 \times 10^{-5} = \frac{[H^+][X^-]}{[HX]}$$

$$= \frac{x^2}{0.100}$$

Thus,

$$[H^+] = 2.0 \times 10^{-3}M$$

and

$$pH = -\log [H^+] = 2.7$$

### Addition of 10.0 ml of NaOH

With this addition, 10/50 of the acid is neutralized and 40/50 remains unneutralized. A buffer exists. Therefore, we can determine the pH as follows:

$$pH = pK - \log \frac{[HA]}{[A^-]}$$

$$= 4.40 - \log \frac{(40/50)}{(10.50)}$$

$$= 3.80$$

### Addition of 20.0 ml of NaOH

$$pH = pK - \log \frac{(30/50)}{(20/50)}$$

$$= 4.22$$

### Addition of 25.0 ml of NaOH

$$pH = pK - \log \frac{(25/50)}{(25/50)}$$

$$= 4.40$$

### Addition of 30.0 ml of NaOH

$$pH = 4.58$$

### Addition of 40.0 ml of NaOH

$$pH = 5.00$$

### Addition of 49.0 ml of NaOH

$$pH = pK - \log \frac{(1/50)}{(49/50)}$$

$$= 4.40 - (-1.69)$$

$$= 6.09$$

### Addition of 50.0 ml of NaOH

The acid is completely neutralized. The solution contains only the salt of a weak base so that the pH is determined from the expression for hydrolysis:

$$A^- + H_2O \rightleftharpoons HA + OH^-$$

$$K_H = \frac{K_W}{K_{HA}} = \frac{[HA][OH^-]}{[A^-]}$$

Substituting and remembering that the concentration of $A^-$ is now one-third the original concentration of HA because of dilution, we obtain:

$$K_H = \frac{K_W}{K_{HA}} = \frac{[HA][OH^-]}{[A^-]}$$

$$[OH^-]^2 = \frac{K_W[A^-]}{K_{HA}}$$

$$= \frac{(1.0 \times 10^{-14})(6.67 \times 10^{-3})}{(4.0 \times 10^{-5})}$$

$$= 1.67 \times 10^{-11}$$

$$[OH^-] = 4.08 \times 10^{-6}$$

Finally,

$$pH = 8.61$$

Addition of 60.0 ml of NaOH

$$[OH^-] = \frac{\text{moles of excess base}}{\text{total volume of solution}}$$

$$= \frac{(0.010 \text{ liter})(0.200M)}{(0.160 \text{ liter})}$$

$$= 0.0125M$$

or

$$pH = 12.10$$

Addition of 100.0 ml of NaOH

$$[OH^-] = \frac{(0.050 \text{ liter})(0.200M)}{(0.200 \text{ liter})}$$

$$= 0.500M$$

$$pH = 12.70$$

SELF-TEST    Complete the test in 40 minutes. The test covers Chapters 16 and 17.

1. How many grams of $Ag^+$ are in 250 ml of a $0.100M$ $K_2CrO_4$ solution that is saturated with $Ag_2CrO_4$? The $K_{SP}$ of $Ag_2CrO_4$ is $1.9 \times 10^{-12}$.

2. What is the pH at the equivalence point of a titration of 50.0 ml of $0.10M$ trimethylacetic acid with $0.10M$ NaOH? The p$K$ of the ionization of trimethylacetic acid is 5.0.

3. What is the pH of the solution described in problem 2 after 30.0 ml of $0.10M$ NaOH is added?

4. What is the percent hydrolysis of a $0.100M$ solution of sodium acetate? The ionization constant of acetic acid is $1.8 \times 10^{-5}$.

# 18

## Metals

OBJECTIVES

I. You should be able to demonstrate your knowledge of the
following terms by defining them, describing them, or
giving specific examples of them:

alkali metals [18.6]
alkaline-earth metals [18.7]
amalgams [18.3]
Bayer method [18.3]
Bessemer process [18.5]
flotation [18.3]
flux [18.4]
gangue [18.2]
Goldschmidt process [18.4]
Hall process [18.4]
Kroll process [18.4]
lanthanide [18.9]
lanthanide contraction [18.8]
matte [18.4]

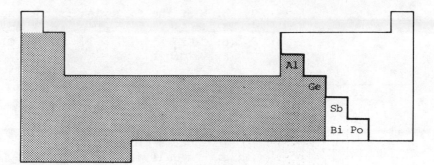

Figure 18.1  Periodic table.

metallurgy [18.3]
Mond process [18.5]
open hearth process [18.5]
ore [18.2]
Parkes process [18.5]
slag [18.4]
smelting [18.4]
thermite process [18.4]
transition metals [18.8]
Van Arkel process [18.5]
zone refining [18.5]

II. You should be familiar with the chemical and physical
properties of the metals shown in the shaded area of the
periodic table in Figure 10.1 of the study guide.

III. You should be familiar with the natural sources of the
most important metals and the methods of refining or
preparing them.

IV. You should be able to apply the basic information of all
previous chapters of your text toward an understanding
of the chemistry of metals.

**EXERCISES**

_____

I. Answer each of the following:

1. Which of the following metals does not have an important oxidation state of 1+?
    (a) Cu              (d) Fe
    (b) Na              (e) Tl
    (c) Au

_____

2. Which of the following is not used as a dessicant?
   (a) NaCl                 (d) $CaSO_4$
   (b) $CaCl_2$             (e) $Ba(ClO_4)_2$
   (c) $Mg(ClO_4)_2$

_____

3. Which of the following is the softest metal?
   (a) Li                   (d) Ba
   (b) Cs                   (e) Na
   (c) Be

_____

4. Aluminum is produced by
   (a) the Hall process     (d) the Bessemer process
   (b) zone refining        (d) the Van Arkel process
   (c) hydrogen reduction

_____

5. The most stable oxidation state of the lanthanide elements is
   (a) 1+                   (d) 5+
   (b) 2+                   (e) 7+
   (c) 3+

_____

6. Which of the following statements is not true of Be with respect to other elements of the group II A?
   (a) It has the highest melting point.
   (b) It is the most dense.
   (c) It has the highest first ionization potential.
   (d) It is the hardest.
   (e) It has the smallest atomic radius.

_____

7. Which of the following transition metals has the greatest number of possible oxidation states?
   (a) Y, yttrium           (d) Zn, zinc
   (b) Sc, scandium         (e) Au, gold
   (c) Ru, ruthenium

_____

8. Which of the following has the smallest atomic radius?
   (a) Mn, manganese        (d) Sc, scandium
   (b) Tc, technetium       (e) Zn, zinc
   (c) Re, rhenium

_____

9. Common impurities in pig iron obtained from a blast furnace include all but one of the following. Identify the element that is not a contaminant.
   (a) carbon               (d) sulfur
   (b) silicon              (e) titanium
   (c) phosphorus

_____

10. Which of the following is commonly found free in nature?
   (a) Fe                   (d) Li
   (b) W                    (e) Au
   (c) Ba

11. Which of the following metals has the lowest melting point?
    (a) Ag                     (d) Pd
    (b) Cd                     (e) Rh
    (c) In

12. Which of the following does not contain an unpaired s electron in the valence shell of the ground state atom?
    (a) Rb, rubidium           (d) Sn, tin
    (b) Cu, copper             (e) Au, gold
    (c) Mo, molybdenum

13. Which of the following compounds is the least soluble in water?
    (a) $Be(OH)_2$             (d) $Ba(OH)_2$
    (b) $Ca(OH)_2$             (e) $Mg(OH)_2$
    (c) $Sr(OH)_2$

14. Which of the following elements has common oxidation states of 2+, 4+, and 7+?
    (a) Mn                     (d) La
    (b) Ni                     (e) Cs
    (c) Pt

15. Which of the following elements has common oxidation states of 1+ and 3+?
    (a) Au                     (d) K
    (b) Cd                     (e) Hg
    (c) Sc

II. Answer the following:

1. Complete the following equation. If no reaction occurs, write *NR* in the space provided for products.

    (a) $Al(s) + Cl_2(g) \rightarrow$

    (b) $CaO(s) + H_2O(l) \rightarrow$

    (c) $ZnS(s) + O_2(g) \xrightarrow{heat}$

    (d) $Ce(s) + Br_2(g) \rightarrow$

    (e) $Ba(s) + H_2O(l) \rightarrow$

    (f) $Ni(s) + CO(g) \xrightarrow{heat}$

    (g) $WO_3(s) + H_2(g) \xrightarrow{heat}$

    (h) $Al_4C_3(s) + H_2O(l) \rightarrow$

    (i) $Tl(s) + H^+(aq) \rightarrow$

    (j) $Ca(s) + Cl_2(g) \rightarrow$

    (k) $Pb(s) + H^+(aq) \rightarrow$

(1) $V(s) + H_2O(l) \rightarrow$

(m) $Na(s) + O_2(g) \rightarrow$

(n) $Al(s) + O_2(g) \xrightarrow{heat}$

(o) $Fe(s) + H^+(aq) \rightarrow$

2. Write the name of each of the following compounds in the space provided:

_____  (a) $Be(OH)_2$

_____  (b) $LiNO_3$

_____  (c) $PbS$

_____  (d) $MoO_3$

_____  (e) $MnO_2$

_____  (f) $Li_2O$

_____  (g) $UF_4$

_____  (h) $Cu_2S$

_____  (i) $TiCl_3$

_____  (j) $Hg_2Cl_2$

## ANSWERS TO EXERCISES

I. Properties of metals

1. Fe
   [18.6, 18.8, 18.10]

Each element of group I A has an oxidation state of 1+. The 1+ oxidation state is an important oxidation state of the transition metals Cu, Au, Ag, and Hg; compounds of $Cu^+$, $Ag^+$, $Au^+$, and $Hg^{2+}$ are common. Thallium is the only group III A metal that has an oxidation state of 1+.

2. NaCl
   [18.7]

The ions of group II A metals and many compounds containing such ions hydrate readily.

3. Cs
   [18.1, 18.6, 18.7]

The softness of the metals of groups I A and II A increases down the group with increasing atomic number. The metals of group II A are harder than the metals of group I A.

4. Hall process
   [18.4]

5. 3+
   [18.9]

6. It is the most dense.
   [18.1]

Both barium and strontium are more dense than beryllium.

7. Ru
   [18.8]

An element of the transition series that contains be-
four and six electrons in the outer $d$ shell can form a
variety of oxidation states.

8. manganese
   [18.8]

Atomic size initially decreases and then increases across
a period of transition elements from left to right; the
minimum is reached near the center of the period. Atomic
size increases with increasing atomic number down a group.

9. titanium
   [18.4]

10. Au [18.2]

11. In [18.1]

12. Sn [18.6, 18.8,
       18.11]

13. Be(OH)$_2$ [18.7]

14. Mn [18.8]

15. Au [18.8]

## II. Reactions of metals and compounds containing metals

1. (a) $2Al(s) + 3Cl_2(g) \rightarrow 2AlCl_3(s)$ [18.10]

   (b) $CaO(s) + H_2O(l) \rightarrow Ca(OH)_2(aq)$ [18.7]

   (c) $2ZnS(s) + 3O_2(g) \xrightarrow{heat} 2ZnO(s) + 2SO_2(g)$ [18.4]

   (d) $2Ce(s) + 3Br_2(g) \rightarrow 2CeBr_3(s)$ [18.9]

   (e) $Ba(s) + 2H_2O(l) \rightarrow Ba(OH)_2(aq) + H_2(g)$ [18.7]

   (f) $Ni(s) + 4CO(g) \xrightarrow{heat} Ni(CO)_4(g)$ [18.5]

   (g) $WO_3(s) + 3H_2(g) \xrightarrow{heat} W(s) + 3H_2O(g)$ [18.4]

   (h) $Al_4C_3(s) + 12H_2O(l) \rightarrow 4Al(OH)_3(s) + 3CH_4(g)$ [18.7]

   (i) $2Tl(s) + 2H^+(aq) \rightarrow 2Tl^+(aq) + H_2(g)$ [18.10]

   (j) $Ca(s) + Cl_2(g) \rightarrow CaCl_2(s)$ [18.7]

   (k) $Pb(s) + 2H^+(aq) \rightarrow Pb^{2+}(aq) + H_2(g)$ [18.11]

   (l) $V(s) + H_2O(l) \rightarrow NR$ [18.8]

   (m) $2Na(s) + O_2(g) \rightarrow Na_2O$ [18.6]

   (n) $4Al(s) + 3O_2(g) \xrightarrow{heat} 2Al_2O_3(s)$ [18.10]

   (o) $Fe(s) + 2H^+(aq) \rightarrow Fe^{2+}(aq) + H_2(g)$ [18.8]

2. (a) beryllium hydroxide [18.7]
   (b) lithium nitrate [18.6]
   (c) lead(II) sulfide [18.11]
   (d) molybdenum(VI) oxide [18.4, 18.8]

(e) manganese(IV) oxide, sometimes called manganese dioxide [18.8]
(f) lithium oxide [18.6]
(g) uranium(IV) fluoride [18.4]
(h) copper(I) sulfide, also called cuprous sulfide [18.4]
(i) titanium(III) chloride [18.8]
(j) mercury(I) chloride, also called mercurous chloride [18.8]

SELF-TEST        Complete the test in 10 minutes:

1. In the spaces provided, write the formulas of the products obtained from the following reactions:

(a) $Mg(s) + H_2O(g) \rightarrow$

(b) $Na(s) + O_2(g) \rightarrow$

(c) $Li(s) + N_2(g) \rightarrow$

(d) $HgS(s) + O_2(g) \xrightarrow{\text{heat}}$

(e) $K(s) + O_2(g) \rightarrow$

(f) $Ba(s) + P_4(l) \xrightarrow{\text{heat}}$

(g) $Ce(s) + S(l) \xrightarrow{\text{heat}}$

# Complex Compounds

OBJECTIVES

I. You should be able to demonstrate your knowledge of the
following terms by defining them, describing them, or
giving specific examples of them:

chelate [19.1]
coordination isomers [19.4]
coordination number [19.1]
crystal field theory [19.5]
degenerate [19.5]
enantiomortph [19.4]
geometric isomers [19.4]
high-spin state [19.5]
hydrate isomers [19.4]
inner complexes [19.5]
ionization isomers [19.4]
isomers [19.4]
labile [19.2]
ligand [19.1]

283

ligand field theory [19.5]
linkage isomers [19.4]
low-spin state [19.5]
optical isomers [19.4]
outer complexes [19.5]
racemic [19.4]
stereoisomers [19.4]
structural isomers [19.4]
valence bond theory [19.5]

II. Given the names or chemical formulas of complex compounds, you should be able to write the corresponding chemical formulas or names.

III. You should be able to draw and identify the various types of isomers.

IV. You should be able to discuss the bonding in complex compounds.

**EXERCISES**

I. Answer each of the following with *true* or *false*. If the statement is false, correct it.

_____

1. In general, $\Delta_O$ is larger for low-spin octahedral complexes of a metal than for high-spin octahedral complexes of that metal.

_____

2. The ethylenediaminetetraacetate ion, EDTA, is a sexadentate chelate.

_____

3. Inner octahedral complexes of iron (III) have all electrons paired.

_____

4. Outer octahedral complexes of iron (III) have five unpaired electrons.

_____

5. Low-spin and high-spin octahedral complexes only exist for $d^4$, $d^5$, $d^6$, and $d^7$ ions.

_____

6. Tetrahedral complexes never exist as *cis-trans* isomers.

_____

7. Tetrahedral complexes never exist as optical isomers.

_____

8. With the addition of $Ag^+$, 2 moles of chloride can be precipitated as AgCl from a solution containing 1 mole of $[Cr(H_2O)_5Cl]Cl_2 \cdot H_2O$.

_____

9. Labile complexes rapidly undergo reactions in which the ligands are replaced.

_____

10. Both chlorophyll and hemoglobin are metal ion complexes of porphyrins.

II. Write the name of each of the following complex compounds in the space provided:

_____ 1. $[Co(NH_3)_6]Cl_3$

_____ 2. $K_4[Fe(CN)_6]$

_____ 3. $[Co(NH_3)_5(SO_4)]Br$

_____ 4. $[Co(NH_3)_5Br]SO_4$

_____ 5. $[Pt(NH_3)_4Cl_2]PtCl_4$

_____ 6. $[Co(H_2O)_6]Cl_2$

_____ 7. $Ir(NH_3)_3Cl_3$

_____ 8. $K_3[Fe(CN)_6]$

_____ 9. $NH_4[Cr(SCN)_4(NO)_2]$

_____ 10. $Pt(NH_3)_2Cl_2$

_____ 11. $[Ni(H_2O)_5Cl]Cl$

_____ 12. $[Cr(NH_3)_6][Cr(CN)_6]$

III. Do the following:

1. Using Figure 19.1, identify the type of isomers represented by each of the following pairs:

_____ (a) $[Co(en)_2ClBr]Cl$ and $[Co(en)_2Cl_2]Br$

_____ (b) $[Pd(dipy)(SCN)_2]$ and $[Pd(dipy)(NCS)_2]$

(c)

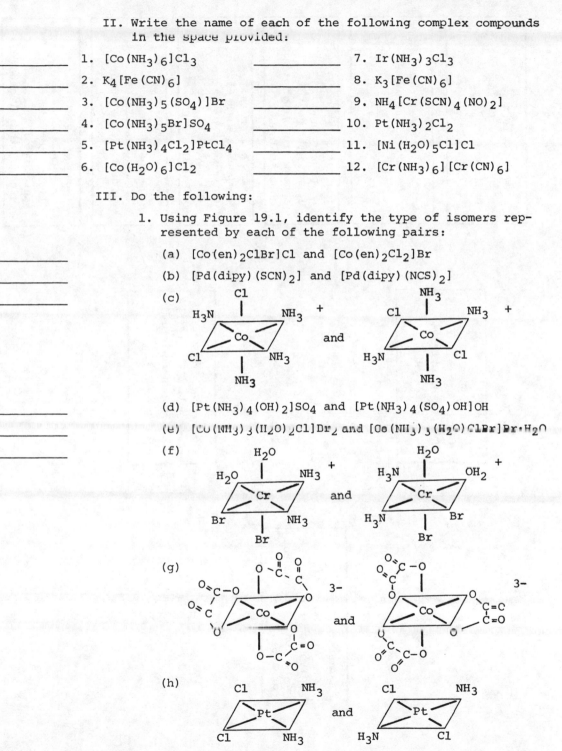

_____ (d) $[Pt(NH_3)_4(OH)_2]SO_4$ and $[Pt(NH_3)_4(SO_4)OH]OH$

_____ (e) $[Co(NH_3)_3(H_2O)_2Cl]Br_2$ and $[Co(NH_3)_3(H_2O)ClBr]Br \cdot H_2O$

(f)

(g)

(h)

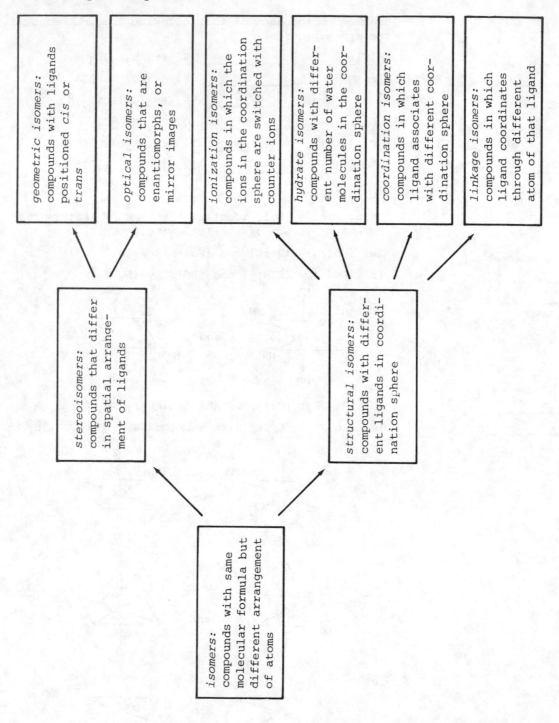

FIGURE 19.1  Isomerism in complex compounds

2. Which of the following complexes cannot have an optical isomer?

(a) *trans*-Pt(NH₃)₂Cl₂, square planar

(b)

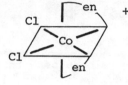

(c)

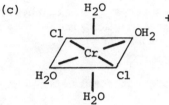

(d)

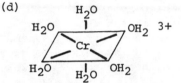

3. Which of the complexes in problem 2 above can exist as *cis-trans* isomers?

4. An elemental analysis of an unknown compound indicates an empirical formula of $CrN_4Cl_3H_{12}$. When dissolved in water, the compound forms two ions, and 1 mole of chloride ion can be precipitated for each mole of compound. The molecular weight of the compound is found to be 226. The compound exists in two isomeric forms. Draw the isomers.

5. The compound *cis*-Pt(NH₃)₂Cl₂ is an active anti-tumor agent. Does it have an optical isomer?

6. A compound when analyzed shows 65.0 percent Pt, 23.6 percent Cl, 9.3 percent N, and 2.0 percent H. The compound has a molecular weight of 600 and forms two ions in solution. No chloride can be removed from solution by the addition of $Ag^+$, and no ammonia can be evolved. Draw two possible isomers of the compound and identify the type of such isomers.

7. Determine whether each of the following transition metal ions is a $d^1$, $d^2$, $d^3$, . . . , $d^9$, or $d^{10}$ ion:

(a) $V^{5+}$

(b) $[Fe(CN)_6]^{4+}$

(c) hexaaquomanganese(III) ion

(d) $[Ni(H_2O)_5Cl]Cl$

(e) $Ni^{3+}$

(f) $NH_4[Cr(SCN)_4(NO)_2]$

IV. Answer each of the following:

1. Using valence bond theory, diagram the electronic arrangement (3d and higher) of $Fe(H_2O)_6^{2+}$, an outer complex, and that of $Fe(CN)_6^{4-}$, an inner complex.

2. Using crystal field theory, diagram the electronic arrangement of iron in each complex ion mentioned in problem 1 above.

## ANSWERS TO EXERCISES

I. Properties of complex compounds

1. True
   [19.5]

For a high-spin octahedral complex the electrons are distributed in the $t_{2g}$ and $e_g$ orbitals such that the number of unpaired electrons is a maximum. Such a distribution occurs when the pairing energy is greater than $\Delta_o$. For a low-spin octahedral complex the pairing energy is less than $\Delta_o$ and thus the lower-energy orbitals are filled completely before the higher-energy orbitals. For example, in an octahedral complex the high-spin state of a $d^6$ metal ion complex is

$$\underline{\uparrow} \qquad \underline{\uparrow} \quad e_g \qquad \uparrow$$
$$\underline{\uparrow\downarrow} \quad \underline{\uparrow} \quad \underline{\uparrow} \quad t_{2g} \qquad \Delta_o \quad \downarrow$$

and the low-spin state of that $d^6$ metal ion complex is

$$\underline{\quad} \qquad \underline{\quad} \quad e_g \qquad \uparrow$$
$$\underline{\uparrow\downarrow} \quad \underline{\uparrow\downarrow} \quad \underline{\uparrow\downarrow} \quad t_{2g} \qquad \Delta_o \quad \downarrow$$

2. True
   [19.1]

3. False
   [19.5]

The Fe(III) ion is a $d^5$ ion. In inner octahedral complexes of Fe(III) there is one unpaired electron:

$$d^2sp^3$$

4. True     In outer octahedral complexes of Fe(III) the electron
   [19.5]   arrangement is

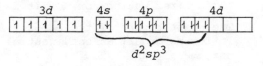

$$d^2sp^3$$

5. True     In an octahedral complex the configuration of a $d^8$ ion is
   [19.5]

$$\underline{\uparrow} \quad \underline{\uparrow} \qquad e_g$$
$$\underline{\uparrow\downarrow} \quad \underline{\uparrow\downarrow} \quad \underline{\uparrow\downarrow} \quad t_{2g}$$
$$d^8$$

and that of a $d^9$ ion is

$$\underline{\uparrow\downarrow} \quad \underline{\uparrow} \qquad e_g$$
$$\underline{\uparrow\downarrow} \quad \underline{\uparrow\downarrow} \quad \underline{\uparrow\downarrow} \quad t_{2g}$$
$$d^9$$

6. True     Each ligand is 109° from each of the other three ligands;
   [19.4]   therefore, *cis* and *trans* isomers cannot exist.

7. False    Optical isomers are possible and have been separated for
   [19.4]   some compounds. For instance, salts of the borosalicyl-
            aldehydro complex cannot be superimposed:

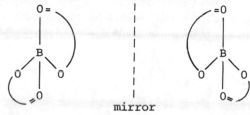

mirror

8. True     The two chlorides outside the brackets are not in the
   [19.4]   first coordination sphere of the metal ion. They exist
            as ions in solution and can be precipitated as AgCl with
            the addition of $Ag^+$ to the solution.

9. True
   [19.2]

10. True
    [19.1]

II. Names of complex compounds

   Refer to Table 19.1 of the study guide or Section 19.3
   of your text if you have any difficulty with this section. After you have checked your answers, write the
   formula of each compound in the space provided.

## TABLE 19.1  Rules for naming complex compounds

1. If the complex compound is a salt, the cation is named first whether or not it is the complex ion.

2. The constituents of the complex are named in the following order: anions, neutral molecules, central metal ion.

3. Anionic ligands are given -o endings; examples are: $OH^-$, hydroxo; $O^{2-}$, oxo; $S^{2-}$, thio; $Cl^-$, chloro; $F^-$, fluoro, $CO_3^{2-}$, carbonato; $CN^-$, cyano; $CNO^-$, cyanato; $C_2O_4^{2-}$, oxalato; $NO_3^-$, nitrato; $NO_2^-$, nitro; $SO_4^{2-}$, sulfato; and $S_2O_3^{2-}$, thiosulfato.

4. The names of neutral ligands are not changed. Exceptions to this rule are: $H_2O$, aquo; $NH_3$, ammine; $CO$, carbonyl; and $NO$, nitrosyl.

5. The number of ligands of a particular type is indicated by a prefix: di-, tri-, tetra-, penta-, and hexa- (for two to six). For complicated ligands (such as ethylenediamine), the prefixes bis, tris-, and tetrakis- (two to four) are employed.

6. The oxidation number of the central ion is indicated by a roman numeral, which is set off by parentheses and placed after the name of the complex.

7. If the complex is an anion, the ending -ate is employed. If the complex is a cation or a neutral molecule, the name is not changed.

1. Hexaaminecobalt(III) chloride                                    _____

2. Potassium hexacyanoferrate(II)                                   _____

3. Sulfatopentaamminecobalt(III) bromide                            _____

4. Bromopentaamminecobalt(III) sulfate                              _____

5. Dichlorotetraammineplatinum(IV) tetrachloroplatinate(II)  _____

6. Hexaaquocobalt(II) chloride                                      _____

7. Trichlorotriammineiridium(III)                                   _____

8. Potassium hexacyanoferrate(III)                                  _____

9. Ammonium tetrathiocyanatodinitrosylchromate(III)          _____

10. Dichloroammineplatinum(II)                                      _____

11. Chloropentaaquonickel(II) chloride                              _____

12. Hexaamminechromium(III) hexacyanochromate(III)                  _____

III. Isomers

See Section 19.3 of your text.

1. (a) ionization
   isomers
   The Cl and Br are switched. In the first compound Br is in the first coordination sphere and in the second compound it is the counter ion.

   (b) linkage
   isomers
   The SCN is bonded to the metal through the S atom in the first compound and through the N atom in the second compound.

   (c) geometric
   isomers
   The chlorines are *cis* in the first compound and *trans* in

   (d) ionization
   isomers

   (e) hydrate
   isomers

   (f) optical
   isomers

   (g) optical
   isomers

   (h) geometric
   isomers

2. (a), (c), and
   (d)
   Mirror images of these complexes can be superimposed on the originals.

3. (a)

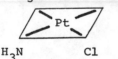

*cis*

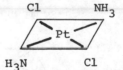

*trans*

   (b)

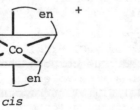

*cis*

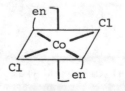

*trans*

(c)

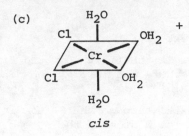

*cis*                                    *trans*

4.

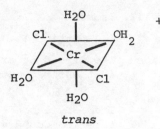

*trans*

The nitrogen—hydrogen ratio indicates the possibility of $NH_3$ in the compound and a formula of $Cr(NH_3)_4Cl_3$ for the compound. Chromium usually forms octahedral complexes. Since 1 mole of chloride ion can be precipitated for each mole of compound, a chloride ion is free in solution; i.e., it is not in the first coordination sphere of Cr. The complex may exist in *cis* and *trans* forms.

*cis*

5. No

The compound

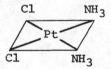

is square planar. Its mirror image can be superimposed on the original.

6. $[Pt(NH_3)_4][PtCl_4]$

    and

$[Pt(NH_3)_3Cl][Pt(NH_3)Cl_3]$

From the data on percent composition an empirical formula of $PtCl_2N_2H_6$, or $Pt(NH_3)_2Cl_2$, can be determined. The compound has a molecular weight of 300; thus, the molecular formula must be $Pt_2(NH_3)_4Cl_4$. The compound is neutral; thus, the charge of the anion must equal the charge of the cation.

7. (a) $d^0$      (d) $d^8$
   (b) $d^6$      (e) $d^7$
   (c) $d^4$      (f) $d^3$

IV. Bonding

1. [19.5]
$Fe(H_2O)_6^{2+}$

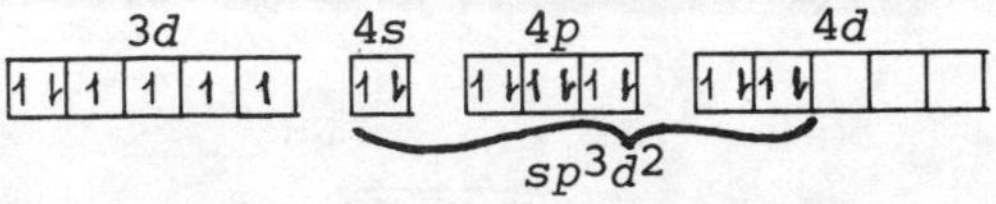

| 3d | 4s | 4p | 4d |
|---|---|---|---|

outer complex

$Fe(CN)_6^{4-}$

| 3d | 4s | 4p | 4d |
|---|---|---|---|

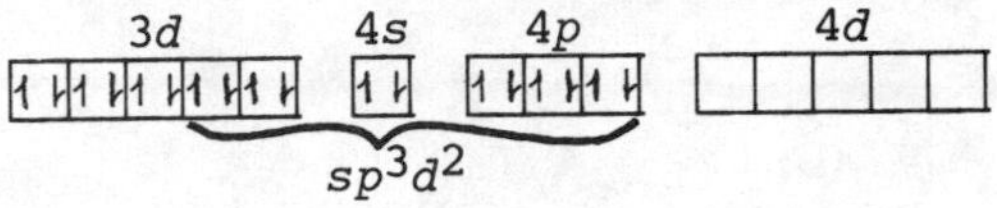

inner complex

2. [19.5]

Iron(II) is a $d^6$ ion. In $Fe(H_2O)_6^{2+}$ the iron(III) is in a high-spin state and in $Fe(CN)_6^{4-}$ it is in a low-spin state:

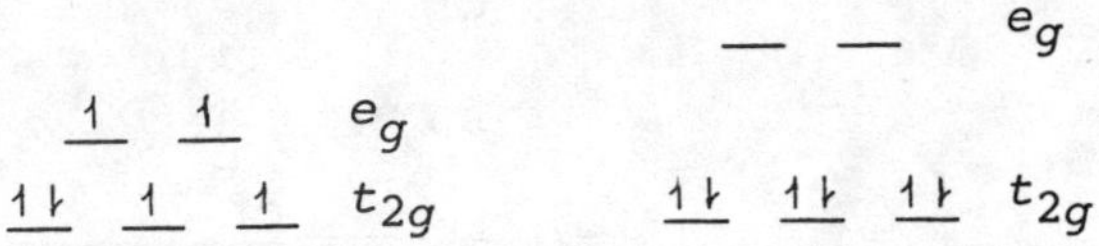

—  —  $e_g$

↑  ↑  $e_g$

↑↓  ↑  ↑  $t_{2g}$

high-spin state
$Fe(H_2O)_6^{2+}$
weak field

↑↓  ↑↓  ↑↓  $t_{2g}$

low-spin state
$Fe(CN)_6^{4-}$
strong field

The crystal field approach indicates the unpaired, nonbonding electrons *and* explains the difference in energy of the d orbitals.

---

## SELF-TEST

Complete the test in 10 minutes:

1. Write the name of each of the following compounds in the space provided:

(a) $K_2[CuCl_4]$

(b) $K_2[PtCl_4]$

(c) $K_4[Fe(CN)_6]$

(d) $Pt(NH_3)_2Cl_4$

(e) $[Co(en)_2(SCN)Cl]Cl$

(f) $[Co(NH_3)_4Br_2]Br$

(g) $[Cu(NH_3)_4]_3[CrCl_6]_2$

2. What is the isomeric relationship between the compounds in each of the following pairs?

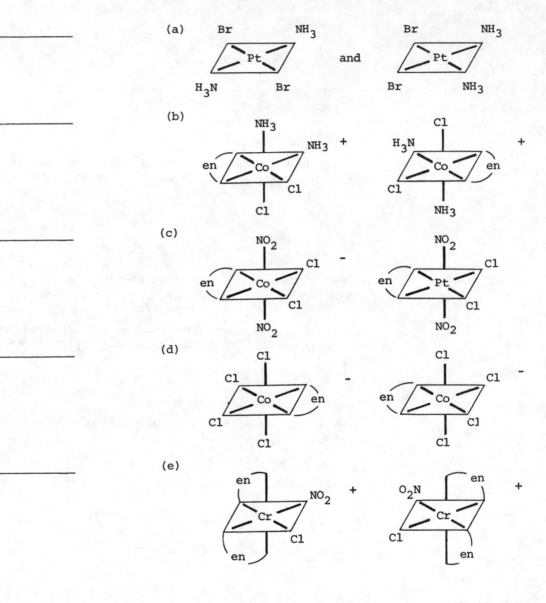

(a)

(b)

(c)

(d)

(e)

# Organic Chemistry

**OBJECTIVES**

I. You should be able to demonstrate your knowledge of the following terms by defining them, describing them, or giving specific examples of them:

alcohols [20.6]
aldehydes [20.7]
alkanes [20.1]
alkenes [20.2]
alkynes [20.3]
amides [20.9]
amines [20.9]
amino acids [20.10]
aromatic [20.4]
carbohydrates [20.11]
carbonium ion [20.5]
carbonyl group [20.7]
carboxyl group [20.8]
carboxylic acids [20.8]

295

chain isomers [20.2]
*cis* [20.2]
conjugated system [20.12]
copolymer [20.12]
esters [20.8]
ethers [20.6]
functional group isomerism [20.6]
geometric isomers [20.2]
homologous [20.1]
ketone [20.7]
Markovnikov's rule [20.5]
*meta-* [20.4]
*normal-* [20.1]
olefins [20.2]
*ortho-* [20.4]
*para-* [20.4]
peptide linkage [20.10]
polymers [20.12]
position isomers [20.2]
primary [20.6, 20.9]
proteins [20.10]
saponification [20.8]
saturated [20.2]
secondary [20.1, 20.6, 20.9]
stereoisomerism [20.2]
structural isomers [20.1]
tertiary [20.1, 20.6, 20.9]
*trans-* [20.2]
unsaturated [20.2]
zwitter ion [20.10]

II. Given the structures or names of simple organic compounds, you should be able to write the corresponding names or structures.

III. You should be familiar with the properties and reactions of the major classes of organic molecules.

IV. You should be able to identify isomers.

**EXERCISES**

_____
_____

I. Write the name of each of the following compounds in the space provided:

1. $CH_3CH_2CH_2CH_3$

2. $CH_3CH_2CHCH_3$
          |
         $CH_3$

3.
$$CH_3$$
$$|$$
$$CH_2$$
$$|$$
$$CH_3-CH_2-C-CH_2-CH_3$$
$$|$$
$$CH_3$$

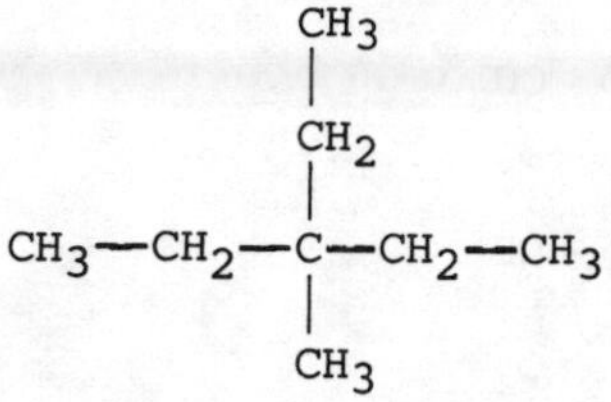

_____

4. 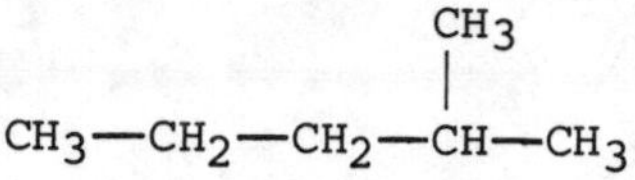
$$CH_3$$
$$|$$
$$CH_3-CH_2-CH_2-CH-CH_3$$

_____

5. 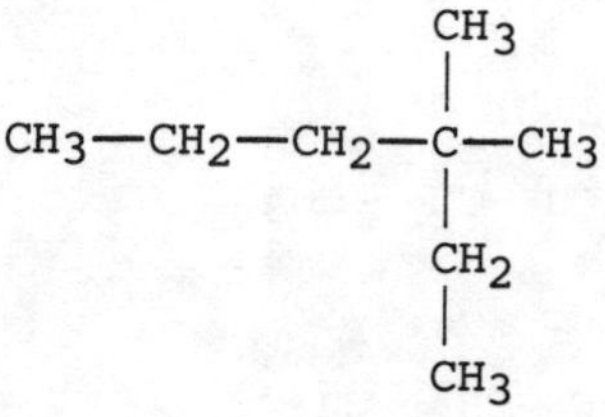
$$CH_3$$
$$|$$
$$CH_3-CH_2-CH_2-C-CH_3$$
$$|$$
$$CH_2$$
$$|$$
$$CH_3$$

_____

6. 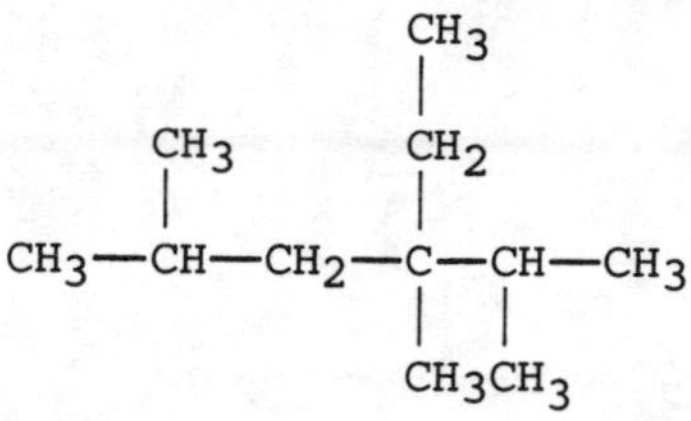
$$CH_3$$
$$|$$
$$CH_3 \quad CH_2$$
$$| \qquad |$$
$$CH_3-CH-CH_2-C-CH-CH_3$$
$$| \quad |$$
$$CH_3 CH_3$$

_____

7. 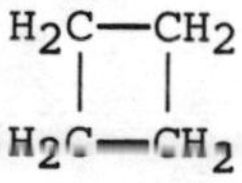
$$H_2C-CH_2$$
$$| \qquad |$$
$$H_2C-CH_2$$

_____

8. 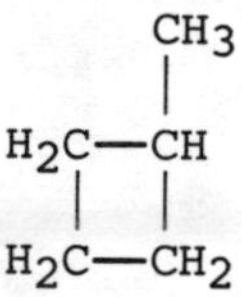
$$CH_3$$
$$|$$
$$H_2C-CH$$
$$| \qquad |$$
$$H_2C-CH_2$$

_____

9. $CH_3-CH=CH-CH_3$

10. $CH_3-CH_2-CH=CH-CH_3$

_____

11. 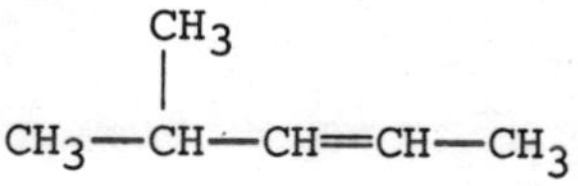
$$CH_3$$
$$|$$
$$CH_3-CH-CH=CH-CH_3$$

_____

12. 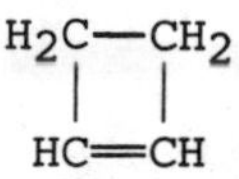
$$H_2C-CH_2$$
$$| \qquad |$$
$$HC=CH$$

_____

13. 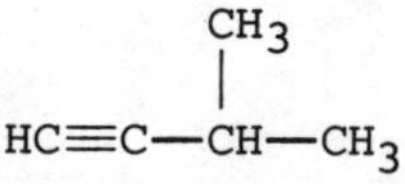
$$CH_3$$
$$|$$
$$HC\equiv C-CH-CH_3$$

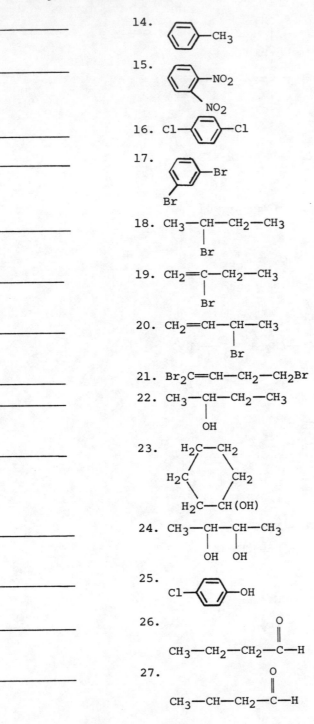

_____    14.

_____    15.

_____    16. Cl—⟨ ⟩—Cl

_____    17.

_____    18. $CH_3$—CH—$CH_2$—$CH_3$
                       |
                       Br

_____    19. $CH_2$=C—$CH_2$—$CH_3$
                       |
                       Br

_____    20. $CH_2$=CH—CH—$CH_3$
                          |
                          Br

_____    21. $Br_2$C=CH—$CH_2$—$CH_2Br$

_____    22. $CH_3$—CH—$CH_2$—$CH_3$
                          |
                          OH

_____    23.

_____    24. $CH_3$—CH—CH—$CH_3$
                          |    |
                          OH   OH

_____    25. Cl—⟨ ⟩—OH

_____    26.
                                 O
                                 ‖
                  $CH_3$—$CH_2$—$CH_2$—C—H

_____    27.
                              O
                              ‖
                  $CH_3$—CH—$CH_2$—C—H

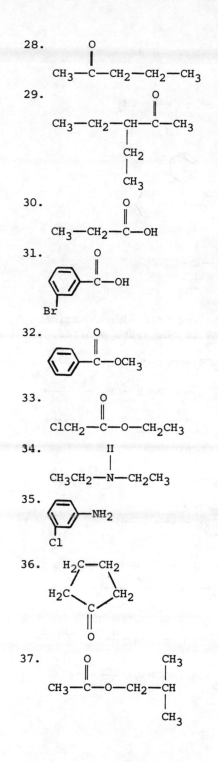

28. 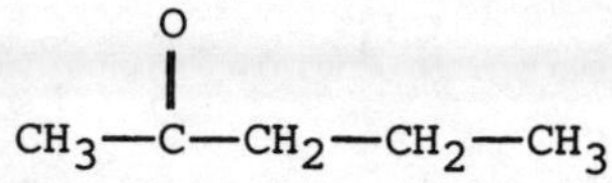

29. 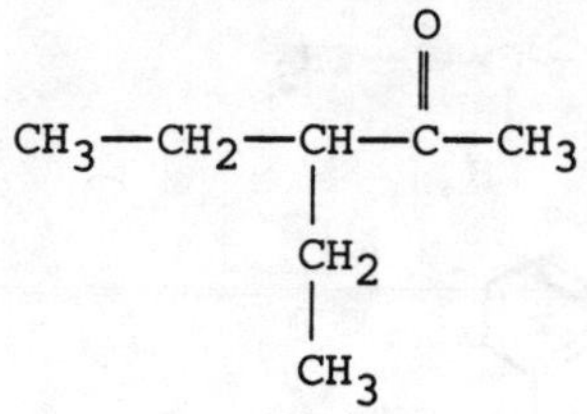

30. 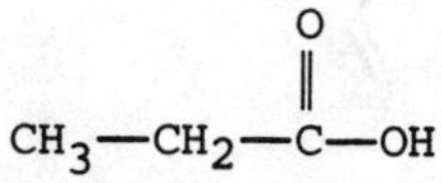

31. 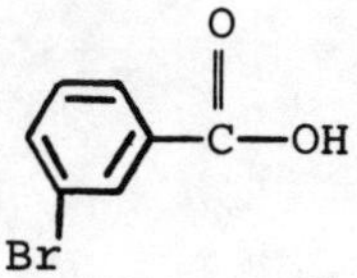

32. 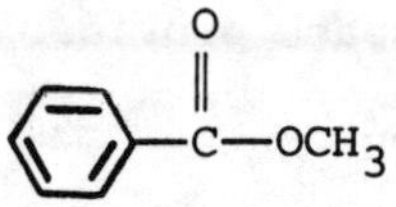

33. 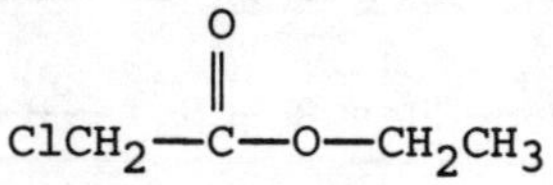

34. 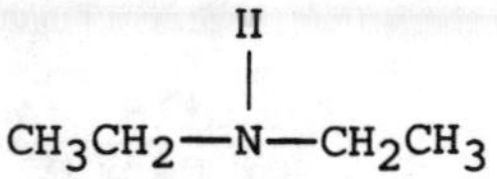

35. 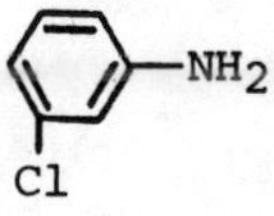

36. 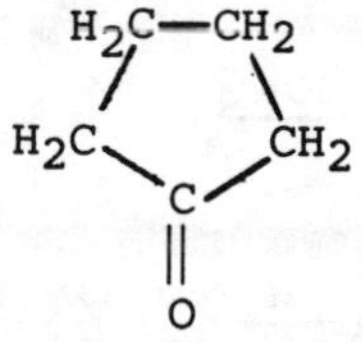

37. 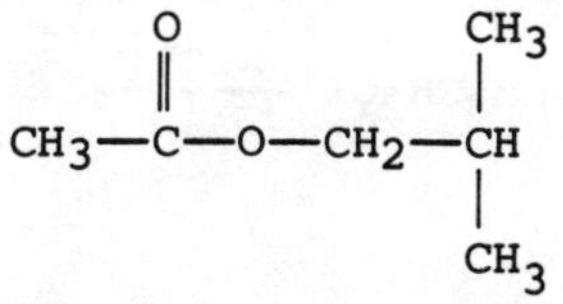

38.

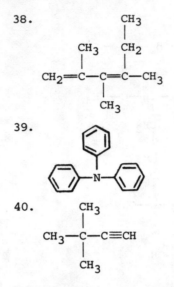

39.

40.

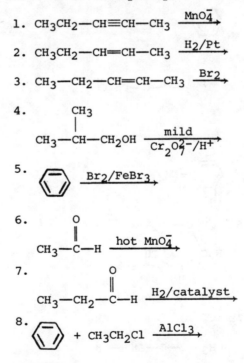

II. Complete the following reactions. If no reaction occurs, write *NR* in the space provided for products.

1. $CH_3CH_2$—CH≡CH—$CH_3$ $\xrightarrow{MnO_4^-}$

2. $CH_3CH_2$—CH=CH—$CH_3$ $\xrightarrow{H_2/Pt}$

3. $CH_3$—$CH_2$—CH=CH—$CH_3$ $\xrightarrow{Br_2}$

4.

$$CH_3-\underset{\overset{|}{CH}}{CH}-CH_2OH \xrightarrow[Cr_2O_7^{2-}/H^+]{mild}$$

with CH₃ on the CH.

5. $\xrightarrow{Br_2/FeBr_3}$

6.

$$CH_3-\overset{\overset{O}{\|}}{C}-H \xrightarrow{hot\ MnO_4^-}$$

7.

$$CH_3-CH_2-\overset{\overset{O}{\|}}{C}-H \xrightarrow{H_2/catalyst}$$

8.

+ $CH_3CH_2Cl$ $\xrightarrow{AlCl_3}$

9.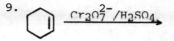

III. Answer each of the following:

   1. The molecule

   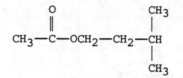

   is an ester that has a banana-like odor. Write a
   reaction for the preparation of the ester from a
   carboxylic acid and an alcohol. Name the carboxylic
   acid, alcohol, and ester.

   2. Teflon is prepared by polymerizing tetrafluoriethene
   (also called tetrafluoroethylene). Draw the structure
   of Teflon.

   3. The compound commonly known as DDT has the following
   structure. Are there any stereoisomers or optical
   isomers of the molecule?

   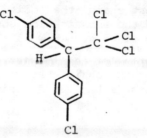

   4. Draw and name all structural isomers with the formula
   $C_7H_{16}$.

   5. Polyvinyl chloride, abbreviated PVC, has the formula

   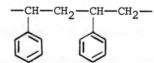

   From what single molecule can it be formed?

   6. Identify the oxidation states of carbon in each compound of the following series:

   $CH_4$, $H_3C$—$OH$, $H_2C$=$O$, $H$—$C\underset{OH}{\overset{O}{\diagup}}$, $O$=$C$=$O$

   Name each compound.

7. In a certain reaction benzaldehyde, which has the formula

and an almond-like aroma, is converted to benzoic acid. Is the reaction an oxidation, a reduction, or a displacement reaction?

8. Draw all possible structural isomers of $C_3H_6O$. Name the isomers. You do not have sufficient information to name all of them.

## ANSWERS TO EXERCISES

1. butane [20.1]

2. methylbutane [20.1]

3. 3-ethyl-3-ethylpentane [20.1]

4. 2-methyl-pentane [20.1]

I. Names of organic compounds

The names of some straight-chain alkanes are given in Table 20.1 of the study guide.

Substituents on the longest chain are given special radical names (see Table 20.2 of the study guide).

When a substituent can be at one of several places on the longest chain, the positions on the chain are numbered and the substitution position is indicated with the appropriate number.

The numbering of the longest chain is always begun at the end of the chain that will give the lowest number to the first substituted position.

TABLE 20.1   Names of Some Straight-Chain Compounds

| Name of Compound | Formula |
| --- | --- |
| methane | $CH_4$ |
| ethane | $C_2H_6$ |
| propane | $C_3H_8$ |
| butane | $C_4H_{10}$ |
| pentane | $C_5H_{12}$ |
| hexane | $C_6H_{14}$ |
| heptane | $C_7H_{16}$ |
| octane | $C_8H_{18}$ |
| nonane | $C_9H_{20}$ |
| decane | $C_{10}H_{22}$ |
| hexadecane | $C_{16}H_{34}$ |
| heptadecane | $C_{17}H_{36}$ |

TABLE 20.2   Names of Simple Radicals

| Formula | Name |
|---|---|
| $CH_3-$ | methyl |
| $CH_3CH_2-$ | ethyl |
| $CH_3CH_2CH_2-$ | *normal*-propyl or *n*-propyl |
| $CH_3CH-$ <br>     $\mid$ <br>     $CH_3$ | isopropyl |
| $CH_3CH_2CH_2CH_2-$ | *normal*-butyl or *n*-butyl |
| $CH_3CHCH_2-$ <br>     $\mid$ <br>     $CH_3$ | isobutyl |
| $CH_3CH_2CH-$ <br>       $\mid$ <br>       $CH_3$ | *secondary*-butyl or *sec*-butyl |
|   $CH_3$ <br>    $\mid$ <br> $CH_3C-$ <br>    $\mid$ <br>   $CH_3$ | *tertiary*-butyl or *tert*-butyl |
| (phenyl ring)— | phenyl |
| $Br-$ | bromo |
| $Cl-$ | chloro |
| $O_2N-$ | nitro |

5. 3,3-dimethyl-
   hexane
   [20.1]

The longest chain is circled and the numbering is included:

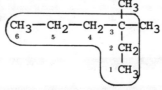

6. 2,3,5-trimethyl-
   3-ethylhexane
   [20.1]

7. cyclobutane
   [20.1]

The names of some cycloalkanes are given in Table 20.3 of the study guide.

TABLE 20.3   Names of Some Cycloalkanes

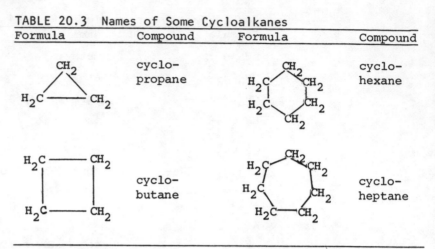

| Formula | Compound | Formula | Compound |
|---|---|---|---|
| | cyclo-propane | | cyclo-hexane |
| | cyclo-butane | | cyclo-heptane |

8. methyl-
   cyclo-
   butane
   [20.1]

The names of the substituents in Table 20.2 of the study guide are used to name such compounds, but the naming becomes more complex when more than one substituent is added because of the possibility of isomers.

9. 2-butene
   [20.2]

The name of an alkene is derived from the name of the corresponding alkane by changing the ending from -*ane* to -*ene*. When necessary, a number is used to show the double-bond position. Because of the restricted rotation about a double bond, *cis* and *trans* isomers exist:

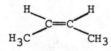  

     *cis*-2-butene              *trans*-2-butene

10. 2-pentene
    [20.2]

The numbering always starts at the end of the chain that gives the lowest numbers to the substituents. Therefore, the compound is 2-pentene, not 3-pentene. For this compound *cis* and *trans* isomers exist:

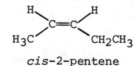  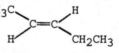

     *cis*-2-pentene             *trans*-2-pentene

11. 4-methyl-
    2-pentene
    [20.2]

For this compound *cis* and *trans* isomers exist:

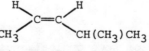

*cis*-4-methyl-2-pentene    *trans*-4-methyl-2-pentene

12. cyclobutene
    [20.2]

The name of a cycloalkene is derived from the name of the corresponding cycloalkane by changing the ending from -ane to -ene.

13. methyl-1-
    butyne
    [20.3]

No number is needed for the methyl group because there is only one possible position. The name of a compound containing a triple bond is derived from the name of the corresponding alkane by changing the ending from -ane to -yne.

14. methylbenzene
    [20.4]

Another acceptable name is toluene.

15. o-dinitro-
    benzene
    [20.4]

The prefixes ortho- (o-), meta- (m-), and para- (p-) are used to designate relative positions of two substituents on a ring.

16. p-dichloro-
    benzene
    [20.4]

17. m-dibromo-
    benzene
    [20.4]

18. 2-bromobutane
    [20.1]

19. 2-bromo-1-
    butene
    [20.2]

20. 3-bromo-1-
    butene
    [20.2]

21. 1,1,4-tri-
    bromo-1-butene
    [20.2]

22. 2-butanol
    [20.6]

A hydrocarbon containing an OH functional group is an alcohol. The name of an alcohol is derived from the name of the corresponding alkane by changing the ending from -ane to -anol. When necessary, a number is used to indicate the position of the OH group.

23. cyclohexanol
    [20.6]

The name of a cycloalcohol is derived from the name of the corresponding cycloalkane by changing the ending -ane to -anol.

24. 2,3-butanediol
    [20.6]

A polyhydroxy alcohol contains more than one OH group. The number of OH groups in a polyhydroxy alcohol is indicated in the name by addition of the appropriate ending to the name of the corresponding alkane: diol, triol, and so on. The position of an OH group is indicated in the name by a number.

25. *p*-chlorophenol
   [20.4, 20.6]

26. butanal          The name of an aldehyde is derived from the name of the
   [20.7]           corresponding alkane by changing the ending -*ane* to -*al*.

27. 3-chlorobutanal  For aldehydes the numbering of carbon atoms always starts
   [20.7]           with the carbon atom that is double bonded to oxygen.

28. 2-pentanone      The name of a ketone is derived from the name of the
   [20.7]           corresponding alkane by changing the ending -*ane* to
                    -*one*. The number of carbon atoms begins at the end of
                    the chain that gives the lowest number of the carbon of
                    the carbonyl group, C=O.

29. 3-ethyl-2-pentanone
   [20.7]

30. propanoic acid   The name of a carboxylic acid is derived from the parent
   [20.8]           hydrocarbon by elision of the final -*e*, addition of the
                    ending -*oic*, and addition of the separate word *acid*. This
                    compound is also called propionic acid.

31. *m*-bromobenzoic acid
   [20.8]

32. methyl benzoate  The name of an ester reflects the alcohol and acid from
   [20.8]           which the ester is derived: the name of the hydrocarbon
                    radical attached to the OH of the alcohol (see Table
                    20.2 of the study guide) is used to indicate the parent
                    alcohol, and the ending -*ate* added to the base of the
                    parent acid is used to indicate the parent acid.

33. ethyl chloroacetate
   [20.8]

34. diethylamine     The name of an amine is formed by the addition of the
   [20.9]           names of the radicals (see Table 20.2 of the study
                    guide) that are attached to the N atom to the word
                    *amine*.

35. *m*-chloroaniline  The molecule ⬡—NH$_2$ is called aniline.

36. cyclopentanone
   [20.7]

37. isopropyl ethanoate
   [20.8]

38. 2,3,4-trimethyl-
   1,3-hexadiene
   [20.2]

39. triphenylamine
    [20.9]

40. 3,3-dimethyl-1-
    butyne
    [20.3]

## II. Reactions of organic compounds

1. [20.8]   $CH_3CH_2$—$CH$=$CH$—$CH_3$ $\xrightarrow[\text{heat}]{\text{MnO}_4^-}$ $CH_3CH_2$—$\overset{\overset{\displaystyle O}{\|}}{C}$—$OH$ + $CH_3$—$\overset{\overset{\displaystyle O}{\|}}{C}$—$OH$
            propanoic acid    ethanoic acid

2. [20.5]   $CH_3CH_2$—$CH$=$CH$—$CH_3$ $\xrightarrow{\text{H}_2/\text{Pt}}$ $CH_3(CH_2)_3CH_3$
            pentane

3. [20.5]   $CH_3$—$CH_2$—$CH$=$CH$—$CH_3$ $\xrightarrow{\text{Br}_2}$ $CH_3$—$CH_2$—$\overset{\overset{\displaystyle Br}{|}}{CH}$—$\overset{\overset{\displaystyle Br}{|}}{CH}$—$CH_3$
            2,3-dibromopentane

4. [20.7]   $CH_3$—$\overset{\overset{\displaystyle CH_3}{|}}{CH}$—$CH_2OH$ $\xrightarrow[\text{Cr}_2\text{O}_7^{2-}/\text{H}^+]{\text{mild}}$ $CH_3$—$\overset{\overset{\displaystyle CH_3}{|}}{CH}$—$\overset{\overset{\displaystyle O}{\|}}{C}$—$H$
            methylpropanal

5. [20.5]   ⬡ $\xrightarrow{\text{Br}_2/\text{FeBr}_3}$ ⬡—Br
            bromobenzene

6. [20.8]   $CH_3$—$\overset{\overset{\displaystyle O}{\|}}{C}$—$H$ $\xrightarrow{\text{hot MnO}_4^-}$ $CH_3$—$\overset{\overset{\displaystyle O}{\|}}{C}$—$OH$
            ethanoic acid
            (acetic acid)

7. [20.7]   $CH_3$—$CH_2$—$\overset{\overset{\displaystyle O}{\|}}{C}$—$H$ $\xrightarrow{\text{H}_2/\text{catalyst}}$ $CH_3$—$CH_2$—$\overset{\overset{\displaystyle OH}{|}}{CH_2}$
            1-propanol

8. [20.5]   ⬡ + $CH_3CH_2Cl$ $\xrightarrow{\text{AlCl}_3}$ ⬡—$CH_2CH_3$ + HCl

9. [20.8]

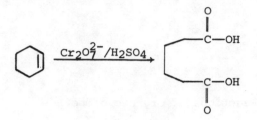

or drawn differently

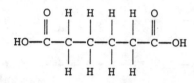

### III. Properties of organic compounds

1. [20.8]

$$CH_3—\overset{\underset{\displaystyle ||}{O}}{C}—OH + HO—CH_2—CH_2—\overset{\underset{\displaystyle CH_3}{|}}{CH} \rightarrow CH_3—\overset{\underset{\displaystyle ||}{O}}{C}—OCH_2—CH_2—\overset{\underset{\displaystyle CH_3}{|}}{CH}$$

ethanoic acid     3-methyl-1-butanol     isopentylacetate
(acetic acid)

2. 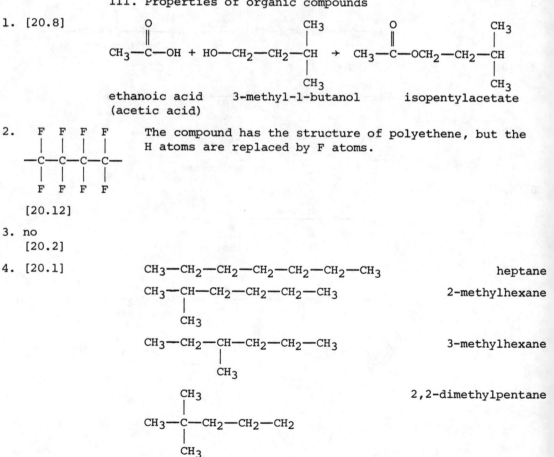    The compound has the structure of polyethene, but the
H atoms are replaced by F atoms.

[20.12]

3. no
[20.2]

4. [20.1]

$CH_3—CH_2—CH_2—CH_2—CH_2—CH_2—CH_3$      heptane

$CH_3—\overset{\underset{\displaystyle CH_3}{|}}{CH}—CH_2—CH_2—CH_2—CH_3$      2-methylhexane

$CH_3—CH_2—\overset{\underset{\displaystyle CH_3}{|}}{CH}—CH_2—CH_2—CH_3$      3-methylhexane

$CH_3—\overset{\overset{\displaystyle CH_3}{|}}{\underset{\underset{\displaystyle CH_3}{|}}{C}}—CH_2—CH_2—CH_2$      2,2-dimethylpentane

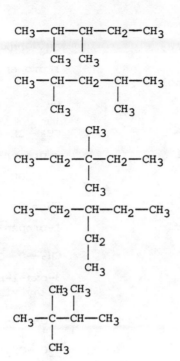

CH3—CH—CH—CH2—CH3                    2,3-dimethylpentane
     |   |
    CH3 CH3

CH3—CH—CH2—CH—CH3                    2,4-dimethylpentane
     |        |
    CH3      CH3

         CH3                  3,3-dimethylpentane
         |
CH3—CH2—C—CH2—CH3
         |
         CH3

CH3—CH2—CH—CH2—CH3                    ethylpentane
         |
         CH2
         |
         CH3

      CH3 CH3                 2,2,3-trimethylbutane
      |   |
CH3—C—CH—CH3
      |
      CH3

5.    HC=CH2

   [20.12]

6. [3.10]    methane  methanol  formaldehyde  formic acid  carbon dioxide

             4-        2-          0           2+          4+

         CH4    H3C—OH    H2C=O          O      C=C=O
                                       H—C
                                         OH

7. oxidation    Note the change in the oxidation state of the carbon
   [20.8]       atom that is double bonded to the oxygen:

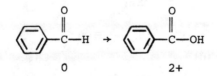

                 0               2+

8. [20.6, 20.7]    You should have been able to name the compounds in the
                    left column, but you may not have been able to name
                    those in the right column:

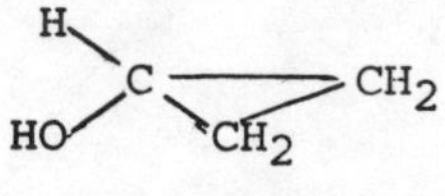

cyclopropanol

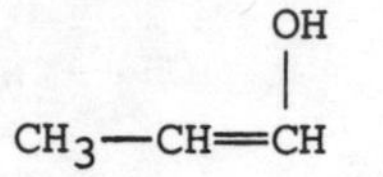

1-propen-1-ol

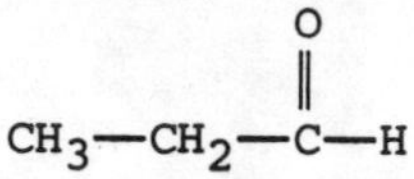

propanal

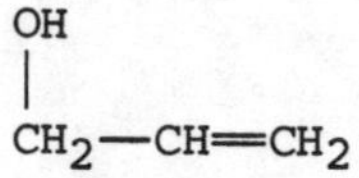

2-propen-1-ol

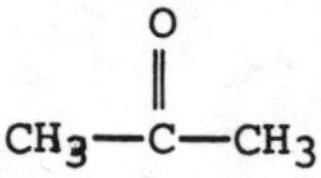

propanone or acetone

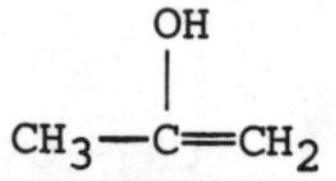

1-propen-2-ol

$CH_3—O—CH{=}CH_2$

3-oxa-1-butene or methylvinyl ether

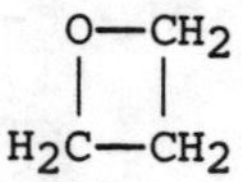

trimethylene oxide

## SELF-TEST

I. Complete the test in 20 minutes:

1. Write the name of each of the following compounds in the space provided:

_______________ (a)

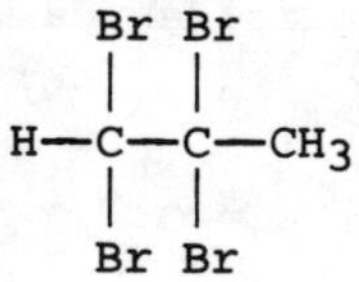

_______________ (b)

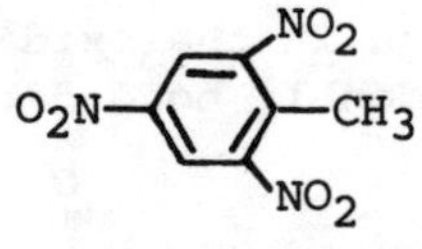

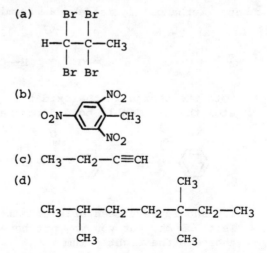

_______________ (c) $CH_3—CH_2—C{\equiv}CH$

_______________ (d)

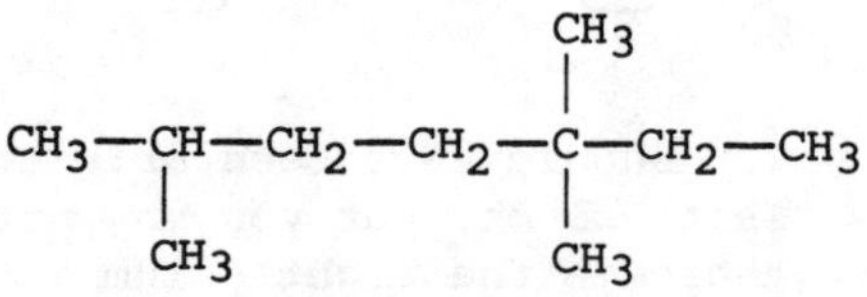

(e)

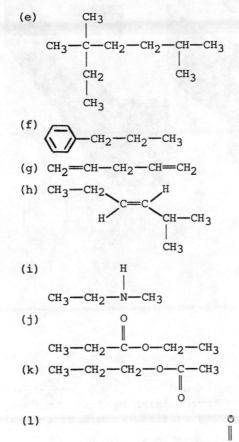

(e)
$$CH_3$$
$$CH_3-C-CH_2-CH_2-CH-CH_3$$
$$CH_2 \quad\quad CH_3$$
$$CH_3$$

(f)
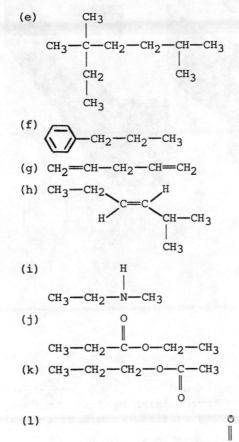
$—CH_2—CH_2—CH_3$

(g) $CH_2{=}CH—CH_2—CH{=}CH_2$

(h) $CH_3—CH_2$   $H$
   $C{=}C$
   $H$   $CH—CH_3$
      $CH_3$

(i)     $H$
   $CH_3—CH_2—N—CH_3$

(j)     $O$
       $\parallel$
   $CH_3—CH_2—C—O—CH_2—CH_3$

(k) $CH_3—CH_2—CH_2—O—C—CH_3$
                $\parallel$
                $O$

(l)             $O$
                $\parallel$
   $CH_3—CH_2—CH_2—CH_2—CH_2—C—H$

2. Write the formulas of the products obtained from each of the following reactions. If no reaction occurs, write *NR*.

   (a) $CH_3CH_2—OH \xrightarrow[\text{heat}]{\text{Cu, O}_2}$

   (b) $CH_2{=}CH_2 \xrightarrow{\text{HCN}}$

   (c) $CH_3—CH_2—C{\equiv}N \xrightarrow[\text{catalyst}]{\text{H}_2}$

3. Draw all structural isomers of $C_3H_6$ and name each.

4. Draw all isomers of $C_3H_8O$ and name each.

# Nuclear Chemistry

OBJECTIVES

I. You should be able to demonstrate your knowledge of the following terms by defining them, describing them, or giving specific examples of them:

alpha particle [21.2]
beta particle [21.2]
binding energy [21.6]
critical mass [21.6]
curie [21.3]
disintegration series [21.4]
electron capture [21.2]
fission [21.6]
fusion [21.6]
gamma radiation [21.2]
Geiger-Müller counter [21.3]
neutron capture [21.5]
nucleons [21.1]

radioactivity [21.2]
radiocarbon dating [21.3]
scintillation counter [21.3]
transuranium elements [21.5]

II. You should be able to write equations for nuclear reactions and account for the mass and energy of such reactions.

III. You should be able to determine the value of the half-life of a radioactive element from experimental data and use the value to determine the amount of radioactive sample present at any time, $t$.

**EXERCISES**

I. Complete and balance the following nuclear reactions:

1. $^{218}_{84}Po \rightarrow {}^{4}_{2}He +$

2. $^{254}_{102}No \rightarrow {}^{4}_{2}He +$

3. $^{238}_{92}U + {}^{1}_{0}n \rightarrow \gamma +$

4. $^{238}_{92}U + {}^{12}_{6}C \rightarrow 6 \; {}^{1}_{0}n +$

5. $^{207}_{84}Po \rightarrow {}^{0}_{1}e +$

6. $^{228}_{88}Ra \rightarrow {}^{0}_{-1}e +$

7. $^{14}_{6}C \rightarrow {}^{0}_{-1}e +$

8. $^{238}_{92}U \rightarrow {}^{4}_{2}He +$

9. $^{87}_{36}Kr \rightarrow {}^{0}_{-1}e +$

10. $^{87}_{36}Kr \rightarrow {}^{1}_{0}n +$

11. $^{13}_{7}N \rightarrow {}^{13}_{6}C +$

12. $^{39}_{19}K + {}^{1}_{0}n \rightarrow 2 \; {}^{1}_{0}n +$

13. $^{238}_{92}U \rightarrow {}^{234}_{90}Th +$

14. $^{235}_{92}U + ^{1}_{0}n \rightarrow ^{139}_{56}Ba + 3\ ^{1}_{0}n +$

15. $^{2}_{1}H + ^{3}_{1}H \rightarrow ^{4}_{2}He +$

16. $^{0}_{-1}e + ^{55}_{26}Fe \rightarrow$

II. Work the following problems:

1. In the electron capture reaction

   $$^{0}_{-1}e + ^{7}_{4}Be \rightarrow ^{7}_{3}Li$$

   how much energy in MeV is released? The mass of $^{7}_{4}Be$ is 7.0169 u and that of $^{7}_{3}Li$ is 7.0160 u.

2. Radon in a tube is often used in cervical cancer therapy. The half-life of radioactive radon is 3.8 days. If 11 micrograms of radon are sealed in a tube, how many micrograms remain after 21 days?

3. How long will it take before the amount of radon in the tube described in problem 2 above is reduced to one millionth of a microgram?

4. The $^{14}_{6}C$ activity of a fiber from an Egyptian mummy shroud is 7.50 disintegrations per minute per gram of carbon. How old is the fiber? The half-life of $^{14}_{6}$ is 5770 years, and the $^{14}_{6}C$ activity of a piece of wo from the outer layer of a freshly cut tree is 15.2 disintegrations per minute per gram of carbon.

5. Use information from problem 4 above to determine th age of the oldest sample that can be dated by the $^{1}_{6}$ technique. Assume that less than one disintegration per minute per gram of carbon cannot be detected with the equipment used with this technique.

6. A sample of $^{147}_{59}Pr$ is prepared and placed in a scintillation counter. The initial counting rate is 200/min. After 36 min the rate is 25 counts/min. What is the half-life of $^{147}_{59}Pr$?

7. How many grams of $^{60}_{27}Co$ will give $75 \times 10^{-3}$ curies, or 75 millicuries, of radiation? The half-life of $^{60}_{27}Co$ is 5.2 years.

**ANSWERS TO EXERCISES**

I. Balancing nuclear equations.

See Section 21.5 of your text.

1. $^{214}_{82}Pb$

An alpha particle is a helium nucleus, $^{4}_{2}He$. Thus, when an alpha particle is emitted, two protons and two neutrons are lost from the nucleus of the atom.

2. $^{250}_{100}Fm$

3. $^{239}_{92}U$

4. $^{244}_{98}Cf$

The capture of a neutron $^{1}_{0}n$, by an atom increases the mass number of that atom by one unit. The loss of a neutron by an atom decreases the mass number of that atom by one unit.

5. $^{207}_{83}Bi$

A positron, $^{0}_{1}e$, is the product of a transformation of a nuclear proton into a nuclear neutron.

6. $^{228}_{89}Ac$

The beta particle, $^{0}_{-1}e$, is a negatively charged, low-mass particle. It is a product of the transformation of a nuclear neutron into a nuclear proton.

7. $^{14}_{7}N$

8. $^{234}_{90}Th$

9. $^{87}_{37}Rb$

10. $^{86}_{36}Kr$

11. $^{0}_{1}e$

12. $^{38}_{19}K$

13. $^{4}_{2}He$

14. $^{94}_{36}Kr$

This is the $^{235}U$ atomic bomb reaction.

15. $^{1}_{0}n$

This is a typical fusion reaction.

16. $^{55}_{25}Mn$

II. Radioactive decay problems

1. 0.8 MeV    We determine the loss of mass:
   [21.2]

$$\text{mass of reactant } {}^{7}_{4}\text{Be} = 7.0169 \text{ u}$$
$$\underline{\text{mass of product } {}^{7}_{3}\text{Li} = 7.0160 \text{ u}}$$
$$\text{loss of mass} = 0.0009 \text{ u}$$

We calculate the energy equivalent of this mass difference by means of Einstein's equation:

$$E = mc^2$$
$$= 0.009 \text{ u } (931 \text{ MeV/u})$$
$$= 0.8 \text{ MeV}$$

Thus, the energy released is 0.8 MeV.

2. 0.24 µg    We first calculate the rate constant for the radio-
   [21.3]    active decay of radon:

$$k = \frac{0.693}{t^{\frac{1}{2}}}$$
$$= \frac{0.693}{3.8 \text{ days}}$$
$$= 0.182 \text{ day}^{-1}$$

We then calculate the fraction of radon remaining after 21 days:

$$\log \left(\frac{N_0}{N}\right) = \frac{kt}{2.303}$$
$$= \frac{(0.182 \text{ day}^{-1})(21 \text{ days})}{2.30}$$
$$= 1.66$$

and

$$\frac{N_0}{N} = 10^{1.66}$$
$$= 10^{0.66} \times 10^{1}$$
$$= 4.6 \times 10^{1}, \text{ or } 46$$

Since $N_0 = 11$ µg, the amount of radon remaining after 21 days is

$$\frac{N_0}{N} = 46$$

$$N = \frac{N_0}{46}$$

$$= \frac{11 \ \mu g}{46}$$

$$= 0.24 \ \mu g$$

**3. 90 days**
**[21.3]**

Since $N_0 = 11 \ \mu g$ and $N = 1 \times 10^{-6} \ \mu g$, the length of time, $t$, can be computed directly:

$$\log \left(\frac{N_0}{N}\right) = \frac{kt}{2.303}$$

$$t = \log \left(\frac{N_0}{N}\right)\left(\frac{2.303}{k}\right)$$

$$= \log \left(\frac{1.1}{1 \times 10^{-6}}\right)\left(\frac{2.3}{0.18 \ \text{day}^{-1}}\right)$$

$$= 9 \times 10^1 \ \text{days, or 90 days}$$

**4. $6.45 \times 10^3$ years**
**[21.3]**

We can determine the value of $k$ from the half-life:

$$k = \frac{0.693}{t^{\frac{1}{2}}}$$

$$= \frac{0.693}{5770 \ \text{years}}$$

$$= 1.20 \times 10^{-4}/\text{year}$$

The number of disintegrations per minuts is proportional to the number of atoms present; therefore, we can substitute the values for the number of disintegrations per minute into the fraction $N_0/N$:

$$\log \left(\frac{N_0}{N}\right) = \frac{kt}{2.303}$$

$$t = \left(\frac{2.303}{k}\right) \log \left(\frac{N_0}{N}\right)$$

$$= \left(\frac{2.303}{1.20 \times 10^{-4}/\text{year}}\right) \log \left(\frac{15.2 \ \text{disint./min}}{7.00 \ \text{disint./min}}\right)$$

$$= 6.45 \times 10^3 \ \text{years}$$

**5. $2 \times 10^4$ years**
**[21.3]**

We determine the time, $t$:

$$\log \left(\frac{N_0}{N}\right) = \frac{kt}{2.303}$$

$$t = \left(\frac{2.303}{k}\right) \log \left(\frac{N_0}{N}\right)$$

$$= \left(\frac{2.3}{1.2 \times 10^{-4}/\text{year}}\right) \log \left(\frac{15 \ \text{disint./min}}{1 \ \text{disint./min}}\right)$$

$$= 2 \times 10^4 \ \text{years}$$

The technique is actually limited to less than 20,000 years by other restrictions.

6. 12 min
   [21.3]

In one half-life, half the original material disintegrates, and the rate drops to 100 counts per minuts. During the second half-life, the rate drops to 50 counts per minute, and during the third half-life it drops to 25 counts per minute. Therefore, 3 half-lives elapse between the initial and the final counts. Thus, the half-life is 1/3 of the elapsed time, or 36 min/3 = 12 min.

7. $6.6 \times 10^{-5}$ g $_{27}^{60}$Co

We convert the activity from millicuries to disintegrations/sec:

$$\text{activity} = (75 \text{ mc}) \left(\frac{3.70 \times 10^7 \text{ disint./sec}}{1 \text{ c}}\right) \left(\frac{1 \text{ c}}{1000 \text{ mc}}\right)$$

$$= 2.78 \times 10^9 \text{ disint./sec, or}$$

$$= 2.78 \times 10^9 \text{ atom/sec}$$

Since the rate constant is expressed in years$^{-1}$, we convert the activity from atoms/sec to atoms/year:

$$? \text{ activity} = \left(\frac{2.78 \times 10^9 \text{ atoms}}{1 \text{ sec}}\right)\left(\frac{60 \text{ sec}}{1 \text{ min}}\right)\left(\frac{60 \text{ min}}{1 \text{ hr}}\right)\left(\frac{24 \text{ hr}}{1 \text{ day}}\right)\left(\frac{365 \text{ days}}{1 \text{ year}}\right)$$

$$= 8.77 \times 10^{16} \text{ atoms/year}$$

We then calculate the rate constant, $k$:

$$k = \frac{0.693}{t^{\frac{1}{2}}}$$

$$= \frac{0.693}{5.2 \text{ years}}$$

$$= 0.133 \text{ year}^{-1}$$

Since activity = $kN$, we can calculate the number of atoms, $N$:

$$\text{activity} = kN$$

$$N = \frac{\text{activity}}{k}$$

$$= \frac{8.77 \times 10^{16} \text{ atoms/year}}{0.133 \text{ year}}$$

$$= 6.59 \times 10^{17} \text{ atoms}$$

Finally we calculate the number of grams:

$$? \text{ g } _{27}^{60}\text{Co} = 6.59 \times 10^{17} \text{ atoms } _{27}^{60}\text{Co} \left(\frac{1 \text{ mol } _{27}^{60}\text{Co}}{6.022 \times 10^{23} \text{ atoms}}\right)\left(\frac{60 \text{ g } _{27}^{60}\text{Co}}{1 \text{ mol } _{27}^{60}\text{Co}}\right)$$

$$= 6.6 \times 10^{-5} \text{ g } _{27}^{60}\text{Co}$$

**SELF-TEST**

Complete the test in 10 minutes:

1. A sample contains $^{35}_{16}S$ as the only radioactive spec-
   cies. How many moles of $^{35}_{16}S$ are in a sample that has
   an activity of $1.01 \times 10^3$ disintegrations per minute.
   The half-life of $^{35}_{16}S$ is 86.6 days.

2. The half-life of $^{31}_{14}Si$ is 2.6 hours. How many grams
   of $^{31}_{14}Si$ must be prepared if $1.0 \times 10^{-12}$ g will be
   needed in an experiment 13 hours later?

3. Calculate the binding energy of an atom of $^{208}_{82}Pb$,
   which has a mass of 208.060 u. The masses of the
   proton, neutron, and electron are 1.007277 u,
   1.008665 u, and 0.0005486 u, respectively.

## ANSWERS TO SELF-TESTS

CHAPTER 1

| | | | |
|---|---|---|---|
| 1. (c) | 5. (d) | 9. (c) | 13. (a) |
| 2. (d) | 6. (a) | 10. (d) | 14. (b) |
| 3. (c) | 7. (c) | 11. (b) | 15. (b) |
| 4. (d) | 8. (b) | 12. (c) | |

CHAPTER 2

I.  1. energy
    2. electron, proton, neutron
    3. neutron
    4. isotopes
    5. 18
    6. diamagnetic
    7. $1s^2 2s^1$, 3

II. 1. The atomic mass unit is defined as one-twelfth the mass of the nuclide $^{12}C$
    2. $n = 5$, $l = 2$, $m = +1$, $s = -\frac{1}{2}$
    3. Ir: $1s^2 2s^2 2p^6 3s^2 3p^6 3d^{10} 4s^2 4p^6 4d^{10} 4f^{14} 5s^2 5p^6 5d^7 6s^2$ and
       $Ir^{2+}$: $1s^2 2s^2 2p^6 3s^2 3p^6 3d^{10} 4s^2 4p^6 4d^{10} 4f^{14} 5s^2 5p^6 5d^7$
    4. 6.94 u

III. 1. (c)    5. (c)    9. (b)    13. (d)
     2. (d)    6. (a)   10. (a)    14. (a)
     3. (d)    7. (a)   11. (d)    15. (b)
     4. (c)    8. (b)   12. (d)

**CHAPTER 3**    I. See the following table:

| Formula of Compound | Name of Compound | Symbol of Element | Oxidation Number of Element |
|---|---|---|---|
| $SnCl_4$ | tin(IV) chloride or stannic chloride | Sn | 4+ |
| $As_2O_3$ | arsenic(III) oxide or arsenous oxide | As | 3+ |
| $Cu(NO_3)_2$ | copper(II) nitrate or cupric nitrate | N | 5+ |
| $MnO_2$ | manganese(IV) oxide | Mn | 4+ |
| $Cr_2O_3$ | chromium(III) oxide | Cr | 3+ |
| $NaC_2H_3O$ | sodium acetate | Na | 1+ |
| $N_2O_4$ | dinitrogen tetroxide | N | 4+ |
| $H_2SO_4$ | sulfuric acid | S | 6+ |
| $NaClO_2$ | sodium chlorite | Cl | 3+ |

II. 1. (d)    4. (d)    7. (c)    9. (a)
    2. (a)    5. (d)    8. (b)   10  (d)
    3. (b)    6. (c)

III.

**CHAPTER 4**

1.

| | Number of Electron Pairs | | Shape of |
|---|---|---|---|
| Formula | Bonding | Nonbonding | Molecule or Ion |
| $SCl_2$ | 2 | 2 | angular |
| $XeF_4$ | 4 | 2 | square planar |
| $AlH_4^-$ | 4 | 0 | tetrahedral |
| $TeCl_4$ | 4 | 1 | trigonal pyramidal or distorted tetrahedral |
| $SeF_5^-$ | 5 | 1 | square pyramidal |

2.

| | Hybrid Orbitals | Geometric Shape |
|---|---|---|
| $PCl_5$ | $dsp^3$ | trigonal bipyramidal |
| $PCl_4^+$ | $sp^3$ | tetrahedral |
| $PCl_6^-$ | $d^2sp^3$ | octahedral |

3.

| | Total Electrons in Orbitals | | | | Bond | Number of Unpaired |
|---|---|---|---|---|---|---|
| Molecule | $\sigma$ | $\sigma*$ | $\pi$ | $\pi*$ | Order | Electrons |
| $Be_2$ | 2 | 2 | 0 | 0 | 0 | 0 |
| $B_2$ | 2 | 2 | 2 | 0 | 1 | 2 |
| $N_2$ | 4 | 2 | 4 | 0 | 3 | 0 |
| $O_2$ | 4 | 2 | 4 | 2 | 2 | 2 |
| $NO^+$ | 4 | 2 | 4 | 0 | 3 | 0 |

4. $\overset{..}{\underset{..}{S}}=C=\overset{..}{\underset{..}{N}}{}^{\ominus} \leftrightarrow :S{\equiv}C-\overset{..}{\underset{..}{N}}{}^{2\ominus}{}^{\oplus} \leftrightarrow :\overset{..}{\underset{..}{S}}{}^{\ominus}-C{\equiv}N:$

**CHAPTER 5**

1. $3.03 \times 10^{22}$ atoms
2. CaO
3. +87.96 kJ/mol
4. $C_6H_5OCl_2$
5. 7.44 g
6. −264.4 kJ

**CHAPTER 6**

I.
1. 16.0 g/mol
2. 105 g/mol
3. 12 mm
4. 800 mm
5. 0.900 g/l
6. 2 liter $NH_3$, 0 liter $O_2$, 8 liter NO, 12 liter $H_2O$
7. 43.9 g/mol

II.
1. $H_2$, 6
2. same, same
3. 1/17
4. intermolecular attractive forces, molecular volume

III.    a. pressure vs. volume

c. absolute temper-
ature vs. volume

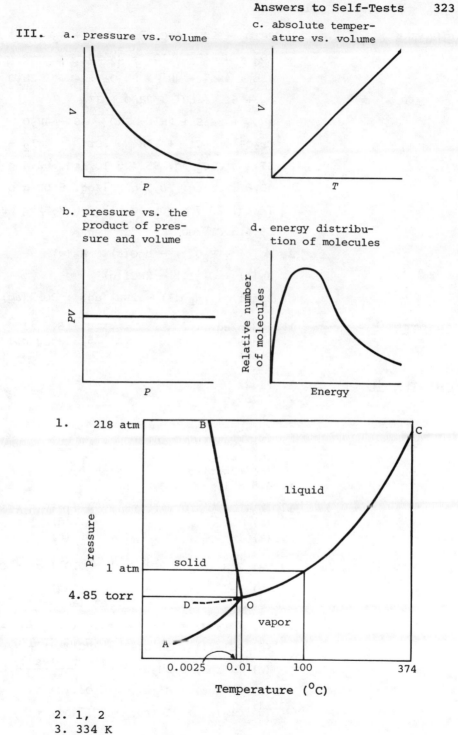

b. pressure vs. the
product of pres-
sure and volume

d. energy distribu-
tion of molecules

CHAPTER 7

1.    218 atm

2. 1, 2
3. 334 K
4. 58.7 g/mol

**CHAPTER 8**

1. $2ClO_2 + 2OH^- \rightarrow ClO_2^- + ClO_3^- + H_2O$

$IO_4^- + H_2AsO_3^- \rightarrow IO_3^- + H_2AsO_4^-$

$4Sb + 4NO_3^- + 4H^+ \rightarrow Sb_4O_6 + 4NO + 2H_2O$

$O_3 + 6I^- + 6H^+ \rightarrow 3H_2O + 3I_2$

$2NO_3^- + 3H_2S + 2H^+ \rightarrow 2NO + 3S + 4H_2O$

$Cr_2O_7^{2-} + 6Cl^- + 14H^+ \rightarrow 2Cr^{3+} + 3Cl_2 + 7H_2O$

2. 67.4 g $ClO_2$/eq, 95.4 g $IO_4^-$/eq, 62.4 g $H_2AsO_3^-$/eq, 40.6 g Sb/eq. 20.7 g $NO_3^-$/eq, 8.00 g $O_3$/eq, 126.9 g $I^-$/eq, 20.7 g $NO_3^-$/eq, 17.0 g $H_2S$/eq, 36.0 g $Cr_2O_7^{2-}$/e 35.4 g $Cl^-$/eq

3. $Mg(s) + H_2O(g) \rightarrow MgO(s) + H_2(g)$

$H_2(g) + Cl_2(g) \rightarrow 2HCl(g)$

$2Na(s) + 2H_2O(l) \rightarrow 2Na^+(aq) + 2OH^-(aq) + H_2(g)$

**CHAPTER 9**

1. 123 g/eq
2. 4.30%
3. 6.00$N$
4. 16.6 ml
5. 125 g/mol

**CHAPTER 10**

I.
1. +1.22 V
2. 0.755 g Mg
3. (a) $Cu^{2+}$
   (b) $Cu^{2+}$/Cu
   (c) positive
   (d) yes
   (e) negative
4. $4.5 \times 10^3$

II.
1. (b)
2. (b)
3. (a)
4. (c)
5. (d)
6. (c)

**CHAPTER 11**

1. (a) $2OH^-(aq) + 2Cl_2(aq) \rightarrow ClO^-(aq) + Cl^-(aq) + H_2O(l$

(b) $FeS(s) + 2H^+(aq) + 2Cl^-(aq)$
$\rightarrow 2Fe^{2+}(aq) + 2Cl^-(aq) + H_2S(g)$

(c) $2Br_2(g) + Ag_2O(s) + H_2O(l) \rightarrow 2AgBr(s) + 2HOBr(ac$

(d) $PCl_3(l) + 3H_2O(l) \rightarrow 3HCl(aq) + H_3PO_3(aq)$

(e) $Cl^-(aq) + 3H_2O(l) \xrightarrow[heat]{electrolysis} ClO_3^-(aq) + 3H_2$

(f) $Cl^-(aq) + H^+(aq) + H_2O(l) \rightarrow NR$

(g) $Cl_2(g) + 2I^-(aq) \rightarrow 2Cl^-(aq) + I_2(s)$

(h) $Ba^{2+}(aq) + SO_4^{2-}(aq) \rightarrow BaSO_4(s)$

(i) $C_{12}H_{22}O_{11}(s) + 11H_2SO_4(conc) \rightarrow 11H_2SO_4 \cdot H_2O + 12C(s)$
   sucrose

   or

   $C_{12}H_{22}O_{11}(s) + 6H_2SO_4 \rightarrow 5H_2SO_4 \cdot 2H_2O + H_2SO_4 \cdot H_2O + 12C$

   or any other balanced equation containing hydrated sulfuric acid species and 12 carbon atoms as products.

(j) $S_2O_3^{2-}(aq) + 2H^+(aq) \rightarrow S(s) + SO_2(g) + H_2O(l)$

**CHAPTER 12**

1.

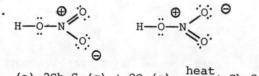

2. (a) $2Sb_2S_3(s) + 9O_2(g) \xrightarrow{heat} Sb_4O_6(g) + 6SO_2(g)$

   (b) $CaO(s) + 3C(s) \xrightarrow{heat} CaC_2(s) + CO(g)$

   (c) $Ca_3N_2(s) + 6H_2O(l) \rightarrow 3Ca^{2+}(aq) + 6OH^-(aq) + 2NH_3(g)$

   (d) $As_4O_6(s) + 6C(s) \xrightarrow{heat} As_4(g) + 6CO(g)$

   (e) $PBr_3(l) + 3H_2O(l) \rightarrow H_3PO_3(aq) + 3HBr(aq)$

   (f) $2PBr_3(l) + O_2(g) \rightarrow 2POBr_3(s)$

   (g) $PI_3(s) + I_2(s) \rightarrow NR$

3. monoprotic acid

**CHAPTER 13**

1. order of $HgCl_2 = 1$, order of $C_2O_4^{2-} = 2$,
   $k = 4.7 \times 10^{-3}M^{-2}$ $min^{-1}$

2. (a) remains the same     (d) decreases
   (b) increases            (e) increases
   (c) remains the same

3. (a) increase             (d) increase
   (b) no change            (e) increase
   (c) no change

4. increase temperature, decrease pressure, remove $SO_2$

5. $K_p = 133$ $atm^{-1}$

6. $6.25 \times 10^{-3}M$

**CHAPTER 14**

1. $-59.5$ kcal/mol [or $-249$ kJ/mol]

2. $K_p = 1.03 \times 10^{16}$

3. reaction is spontaneous, $K_p = 9.84 \times 10^4$

CHAPTER 15

1. $NH_3$, $H^-$
2. $H_2S$
3. $SOCl_2$ ($SO^{2+}$ is the acid ion)
4. $F^-$
5. $\ddot{S}$:
6. $HS^-$, $HC_2H_3O_2$
7. $HPO_4^{2-}$

CHAPTER 16

1. 12.9
2. $2.5 \times 10^{-10}$
3. $2.0 \times 10^{-6}$
4. $0.18M$

CHAPTER 17

1. $1.2 \times 10^{-4}$ g $Ag^+$/250 ml
2. 8.8
3. 5.2
4. 0.0075%

CHAPTER 18

1. (a) $MgO(s)$ $H_2(g)$
   (b) $Na_2O_2(s)$
   (c) $Li_3N(s)$
   (d) $Hg(g)$, $SO_2(g)$
   (e) $KO_2(s)$
   (f) $Ba_3P_2(s)$
   (g) $Ce_2S_3(s)$

CHAPTER 19

1. (a) potassium tetrachlorocuprate(II)
   (b) potassium tetrachloroplatinate(II)
   (c) potassium hexacyanoferrate(II)
   (d) tetrachlorodiammineplatinum(IV)
   (e) chlorothiocyantobis(ethylenediamine)cobalt(III) chloride
   (f) dibromotetraamminecobalt(III) bromide
   (g) tetraamminecopper(II) hexachlorochromate(III)
2. (a) geometric isomers
   (b) same compound
   (c) not isomers, different central metal atom
   (d) same compound
   (e) optical isomers

CHAPTER 20

I. 1. (a) 1,1,2,2-tetrabromopropane
      (b) 1,3,5-trinitrotoluene
      (c) 1-butyne
      (d) 2,5,5-trimethylheptane
      (e) 2,5,5-trimethylheptane
      (f) *n*-propylbenzene
      (g) 1,4-pentadiene
      (h) *trans*-2-methyl-3-hexene
      (i) methylethylamine
      (j) ethyl propionate
      (k) *n*-propyl acetate
      (l) hexanal

   2. (a)
$$CH_3-\overset{\displaystyle O}{\overset{\displaystyle \|}{C}}-H$$

      (b) $CH_3-CH_2-C\equiv N$
      (c) $CH_3-CH_2-CH_2-NH_2$

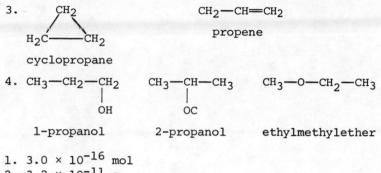

3. cyclopropane

propene

4. 1-propanol    2-propanol    ethylmethylether

CHAPTER 21

1. $3.0 \times 10^{-16}$ mol
2. $3.2 \times 10^{-11}$ g
3. 7.84 MeV